Self-healing Materials

Edited by
Swapan Kumar Ghosh

Further Reading

Schmid, Günter / Krug, Harald / Waser,
Rainer / Vogel, Viola / Fuchs, Harald /Grätzel,
Michael / Kalyanasundaram, Kuppuswamy /
Chi, Lifeng (eds.)

Nanotechnology

9 Volumes

2009
ISBN: 978–3–527–31723–3

Allcock, Harry R.

Introduction to Materials Chemistry

2008
ISBN: 978–0–470–29333–1

Krenkel, Walter (ed.)

Ceramic Matrix Composites

Fiber Reinforced Ceramics and their Applications

2008
ISBN: 978–3–527–31361–7

Öchsner, A., Murch, G., de Lemos, M. J. S.
(eds.)

Cellular and Porous Materials

Thermal Properties Simulation and Prediction

2008
ISBN: 978–3–527–31938–1

Vollath, D.

Nanomaterials

An Introduction to Synthesis, Properties and Applications

2008
ISBN: 978–3–527–31531–4

Ghosh, S. K. (ed.)

Functional Coatings

by Polymer Microencapsulation

2006
ISBN: 978–3–527–31296–2

Butt, Hans-Jürgen / Graf, Karlheinz / Kappl,
Michael

Physics and Chemistry of Interfaces

2006
ISBN: 978–3–527–40629–6

Self-healing Materials

Fundamentals, Design Strategies, and Applications

Edited by
Swapan Kumar Ghosh

WILEY-VCH Verlag GmbH & Co. KGaA

The Editor

Dr. Swapan Kumar Ghosh
ProCoat India Private Limited
Kalayaninagar
Pune-411 014
India

Library of Congress Card No.:
applied for

British Library Cataloguing-in-Publication Data
A catalogue record for this book is available
from the British Library.

**Bibliographic information published by
the Deutsche Nationalbibliothek**
Die Deutsche Nationalbibliothek lists this
publication in the Deutsche Nationalbibli-
ografie; detailed bibliographic data are available in the Internet at <http://dnb.d-nb.de>.

Composition   Laserwords Private Limited,
Chennai, India
Printing   Strauss GmbH, Mörlenbach
Bookbinding   Litges & Dopf GmbH,
Heppenheim
Cover Design   Adam Design, Weinheim

Printed in the Federal Republic of Germany
Printed on acid-free paper

ISBN: 978-3-527-31829-2

Contents

Self-healing Materials: Fundamentals, Design Strategies, and Applications. Edited by Swapan Kumar Ghosh
Copyright © 2009 WILEY-VCH Verlag GmbH & Co. KGaA, Weinheim
ISBN: 978-3-527-31829-2

Preface

Scientists have altered the properties of materials such as metals, alloys, polymers, and so on, to suit the ever changing needs of our society. As we entered into the twenty-first century, search of advanced materials with crack avoidance and long-term durability is on high priority. The challenge for material scientists is therefore to develop new technologies that can produce novel materials with increased safety, extended lifetime and no aftercare or a very less amount of repairing costs. To stimulate this interdisciplinary research in materials technology, the idea of compiling a book came to my mind in 2005. When I contacted one of the pioneer scientists in this field he remarked that it is too early to write a book on such a topic. His opinion was right because the field of material science and technology is rapidly advancing and it would be worth to wait few more years to include the latest updates. Thus this book is complied when the field of self-healing materials research is not matured enough as it is in its childhood.

The title Self-healing Materials itself describes the context of this book. It intends to provide its readers an upto date introduction of the field of self-healing materials (broadly divided into four classes—metals, polymers, ceramics/concretes, and coatings) with the emphasis on synthesis, structure, property, and possible applications. Though this book is mainly devoted to the scientists and engineers in industry and academia as its principle audience, it can also be recommended for graduate courses.

This book with its nine chapters written by international experts gives a wide coverage of many rapidly advancing fields of material science and engineering. The introductory chapter addresses the definition, broad spectrum of strategies, and application potentials of self-healing materials. Chapter 2 summarizes the recent advances in crack healing of polymers and polymer composites. Self-healing in most common polymeric structures occurs through chemical reactions. However, in the case of ionic polymers or ionomers healing follows a different mechanism. This is the subject of Chapter 3. Corrosion causes severe damages to metals. Encapsulated corrosion inhibitors can be incorporated into coatings to provide self-healing capabilities in corrosion prevention of metallic substrates. This is dealt in Chapter 4. Ceramics are emerging as key materials for structural applications. Chapter 5 describes the self-healing capability of ceramic materials. Concrete is the

Self-healing Materials: Fundamentals, Design Strategies, and Applications. Edited by Swapan Kumar Ghosh
Copyright © 2009 WILEY-VCH Verlag GmbH & Co. KGaA, Weinheim
ISBN: 978-3-527-31829-2

most widely used man made materials for structural applications. The possibility of introducing self-healing function in cements is the key subject of Chapter 6. Self-healing in metals is dealt in Chapter 7 while its subsequent Chapter 8 provides an insight of self-healing phenomenon in metallic alloys. The last chapter of this book describes the developments of a model to predict the effects of distributed damages and its subsequent self-healing processes in fiber reinforced polymer composites.

I hope the above mentioned chapters will deliver the readers useful information on self-healing material developments. I am grateful to the contributing authors of this book for their assistance to make this project a success. I would also like to thank the whole Wiley-VCH team involved in this project. Though, last but not least, I would like to dedicate this book to my wife Anjana and son Subhojit for their constant support and encouragement in this venture.

Swapan Kumar Ghosh
September 2008

List of Contributors

Kotoji Ando
Yokohama National University
Department of Energy & Safety Engineering
79-5, Tokiwadai, Hodogaya-ku
Yokohama 240-8501
Japan

Ever J. Barbero
West Virginia University
Mechanical and Aerospace Engineering
Morgantown, WV 26506-6106
USA

Kevin J. Ford
West Virginia University
Mechanical and Aerospace Engineering
Morgantown, WV 26506-6106
USA

Christopher Joseph
Cardiff School of Engineering
Queen's Buildings
The Parade
Newport Road
Cardiff CF24 3AA
United Kingdom

Stephen James Kalista, Jr.
Washington and Lee University
Department of Physics and Engineering
204 West Washington Street
Lexington, VA 24450
USA

Swapan Kumar Ghosh
ProCoat India Private Limited
Kalayaninagar, Pune-411 014
India

Michele V. Manuel
University of Florida
Department of Materials Science and
Engineering
152 Rhines Hall
P.O. Box 116400
Gainesville, FL 32611-6400
USA

Joan A. Mayugo
Escola Politècnica Superior
University de Girona
Campus Montilvi, 17071 Girona
Spain

Wataru Nakao
Yokohama National University
Interdisciplinary Research Center
79-5, Tokiwadai, Hodogaya-ku,
Yokohama, 240-8501,
Japan

Min Zhi Rong
Materials Science Institute
Zhongshan University
135# Xin-Gang-Xi Rd.
Guangzhou 510275
P. R. China

Self-healing Materials: Fundamentals, Design Strategies, and Applications. Edited by Swapan Kumar Ghosh
Copyright © 2009 WILEY-VCH Verlag GmbH & Co. KGaA, Weinheim
ISBN: 978-3-527-31829-2

Erik Schlangen
Delft University of Technology
Department of Civil Engineering and
Geosciences
P.O. Box 5048
2600 GA Delft
The Netherlands

Norio Shinya
Innovative Materials Engineering Laboratory,
Sengen Site,
National Institute for Materials Science
1-2-1, Sengen,
Tsukuba, Ibaraki 305-0047
Japan

Koji Takahashi
Yokohama National University
Division of Materials Science and
Engineering
79-5, Tokiwadai, Hodogaya-ku
Yokohama, 240-8501
Japan

Tao Yin
Materials Science Institute
Zhongshan University
135# Xin-Gang-Xi Rd.
Guangzhou 510275
P. R. China

Ming Qiu Zhang
Materials Science Institute
Zhongshan University
135# Xin-Gang-Xi Rd.
Guangzhou 510275
P. R. China

Mikhail Zheludkevich
Department of Ceramics and Glass
Engineering, CICECO, University of Aveiro,
Campus Universitario de Santiago,
3810-193
Aveiro
Portugal

1
Self-healing Materials: Fundamentals, Design Strategies, and Applications

Swapan Kumar Ghosh

1.1
Introduction

Self-healing materials are no more an illusion and we are not far away from the days when manmade materials can restore their structural integrity in case of a failure. For example, the cracks in buildings can close on their own or the scratches on car bodies can recover their original shiny appearance by itself. Indeed, this is what everyone can see in case of the natural healing of wounds and cuts in living species. Virtually, all materials are susceptible to natural or artificial degradation and deteriorate with time. In the case of structural materials the long-time degradation process leads to microcracks that causes a failure. Thus, repairing is indispensable to enhance reliability and lifetime of materials. Though scientists are inspired by the natural process of blood clotting or repairing of fractured bones, incorporating the same concept into engineering materials is far from reality due to the complex nature of the healing processes in human bodies or other animals [1–6]. However, the recent announcement from Nissan on the commercial release of scratch healing paints for use on car bodies has gained public interest on such a wonderful property of materials [7].

1.2
Definition of Self-healing

Self-healing can be defined as the ability of a material to heal (recover/repair) damages automatically and autonomously, that is, without any external intervention. Many common terms such as self-repairing, autonomic-healing, and autonomic-repairing are used to define such a property in materials. Incorporation of self-healing properties in manmade materials very often cannot perform the self-healing action without an external trigger. Thus, self-healing can be of the following two types:

- autonomic (without any intervention);

Self-healing Materials: Fundamentals, Design Strategies, and Applications. Edited by Swapan Kumar Ghosh
Copyright © 2009 WILEY-VCH Verlag GmbH & Co. KGaA, Weinheim
ISBN: 978-3-527-31829-2

- nonautonomic (needs human intervention/external triggering).

Here, in this review, different types of healing processes are considered as self-healing in general. Currently, self-healing is only considered as the recovery of mechanical strength through crack healing. However, there are other examples where not only the cracks but also small pinholes can be filled and healed to have better performance. Thus, this review addresses recovery of different types of properties, of materials.

1.3
Design Strategies

The different types of materials such as plastics/polymers, paints/coatings, metals/alloys, and ceramics/concrete have their own self-healing mechanisms. In this chapter, different types of self-healing processes are discussed with respect to design strategies and not with respect to types of materials and their related self-healing mechanisms as they are considered in the other chapters of this book. The different strategies of designing self-healing materials are as follows:

- release of healing agent
- reversible cross-links
- miscellaneous technologies
  - electrohydrodynamics
  - conductivity
  - shape memory effect
  - nanoparticle migration
  - co-deposition.

1.3.1
Release of Healing Agents

Liquid active agents such as monomers, dyes, catalysts and hardeners containing microcapsules, hollow fibers, or channels are embedded into polymeric systems during its manufacturing stage. In the case of a crack, these reservoirs are ruptured and the reactive agents are poured into the cracks by capillary force where it solidifies in the presence of predispersed catalysts and heals the crack. The propagation of cracks is the major driving force of this process. On the other hand, it requires the stress from the crack to be relieved, which is a major drawback of this process. As this process does not need a manual or external intervention, it is autonomic. The following sections give an overview of different possibilities to explore this concept of designing self-healing materials.

1.3.1.1 Microcapsule Embedment

Microencapsulation is a process of enclosing micron-sized particles of solids, droplets of liquids, or gases in an inert shell, which in turn isolates and protects them from the external environments [8–11]. The inertness is related to the reactivity of the shell to the core material. The end product of the microencapsulation process is termed as *microcapsules*. It has two parts, namely, the core and the shell. They may have spherical or irregular shapes and may vary in size ranging from nano- to microscale. Healing agents or catalysts containing microcapsules are used to design self-healing polymer composites. Early literature [12, 13] suggests the use of microencapsulated healing agents in a polyester matrix to achieve a self-healing effect. But they were unsuccessful in producing practical self-healing materials. The first practical demonstration of self-healing materials was performed in 2001 by Prof. Scot White and his collaborators [14]. Self-healing capabilities were achieved by embedding encapsulated healing agents into polymer matrix containing dispersed catalysts. The self-healing strategy used by them is shown in Figure 1.1.

In their work, they used dicyclopentadiene (DCPD) as the liquid healing agent and Grubbs' catalyst [bis(tricyclohexylphosphine) benzylidine ruthenium (IV) dichloride] as an internal chemical trigger and dispersed them in an epoxy matrix. The monomer is relatively less expensive and has high longevity and low viscosity. Figure 1.2 shows a representative morphology of encapsulated DCPD and Grubb's catalyst [15–18].

When DCPD comes into contact with the Grubbs' catalyst dispersed in the epoxy resin a *ring opening metathesis polymerization* (ROMP) [19, 20] starts and a highly cross-linked tough polycyclopendiene is formed that seals the crack (Figure 1.3).

The low viscosity of the monomer helps it to flow into the crack plane. The authors have demonstrated that as much as 75% of the recovery of fracture toughness compared to the original specimen can be achieved [17]. The same authors later used encapsulated catalyst instead of encapsulated monomer healing agent [21]. Monomers such as hydroxyl-functionalized polydimethylsiloxane (HOPDMS) and polydiethyoxysilane (PDES) were added to vinyl ester matrix where they stay as microphase-separated droplets. The polyurethane microcapsules containing the catalyst di-*n*-dibutyltin dilaurate (DBTL) is then dispersed in the matrix. Upon rupture of these capsules the catalyst reacts with the monomer and polycondensantaion reaction of the monomers takes place. Keller *et al.* [22] have designed polydimethylsiloxane (PDMS)-based self-healing elastomers using two different types of microcapsules, namely, a resin capsule and an initiator capsule. The size of microcapsules on the self-healing efficiency was also investigated by White *et al.* [23].

Recently, White *et al.* has reported the synthesis of self-healing polymer composites without the use of catalysts [24]. Following these reports [25–30], a large number of research groups around the globe have involved actively in this radical field. Yin *et al.* recently reported the use of a latent curing agent, $CuBr_2(2\text{-MeIm})_4$, instead of

Polymer matrix · Propagating crack

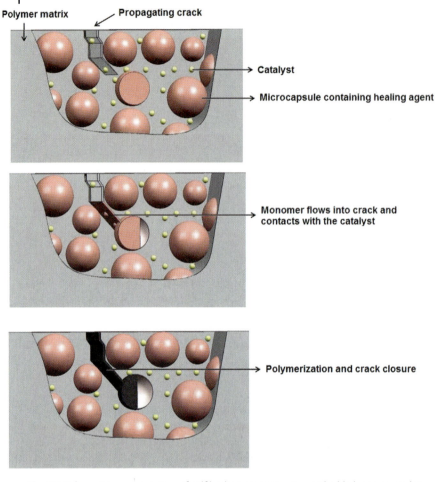

Catalyst

Microcapsule containing healing agent

Monomer flows into crack and contacts with the catalyst

Polymerization and crack closure

Fig. 1.1 Schematic representation of self-healing concept using embedded microcapsules.

solid phase catalyst, to design self-healing materials using ROMP reactions [31]. More detailed discussion on self-healing polymer composites designed through healing agent-based strategy can be found in Chapter 2 of this book.

The critical factors that influence the microencapsulation-based self-healing approach to produce an effective self-healing material are summarized in Table 1.1.

1.3.1.2 Hollow Fiber Embedment

Microcapsule-based self-healing approach has the major disadvantage of uncertainty in achieving complete and/or multiple healing as it has limited amount of healing agent and it is not known when the healing agent will be consumed entirely. Multiple healing is only feasible when excess healing agent is available in the matrix after the first healing has occurred. Thus, to achieve multiple healing in composite materials, another type of reservoir that might be able to deliver larger

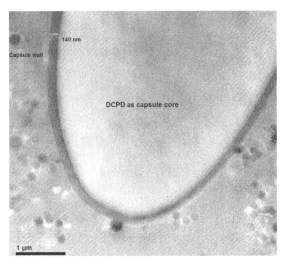

Fig. 1.2 Light microscopic picture of encapsulated DCPD and Grubb's catalyst. (Reprinted with permission from [18].)

amount of liquid healing agent was developed by Dry and coworkers [12, 32, 33]. However, they have achieved only limited success using their approach. Later, large diameter capillaries were embedded into resins by Motuku *et al.*, but the trials were unsuccessful as well [34]. Belay *et al.* have used smaller hollow glass fibers (Hollex fibers) filled with resin [35]. Composites system formulated on the basis of these filled glass fibers were unable to deliver the resin into the crack owing to the use of high viscous epoxy resins, and curing was also not good.

Bond and coworkers later developed a process to optimize the production of hollow glass fibers [36] and used these fibers as the container for liquid healing agents and/or dyes [37–41]. These borosilicate glass fibers' have diameter ranging from 30 to 100 µm with hollowness of 55% (Figure 1.4).

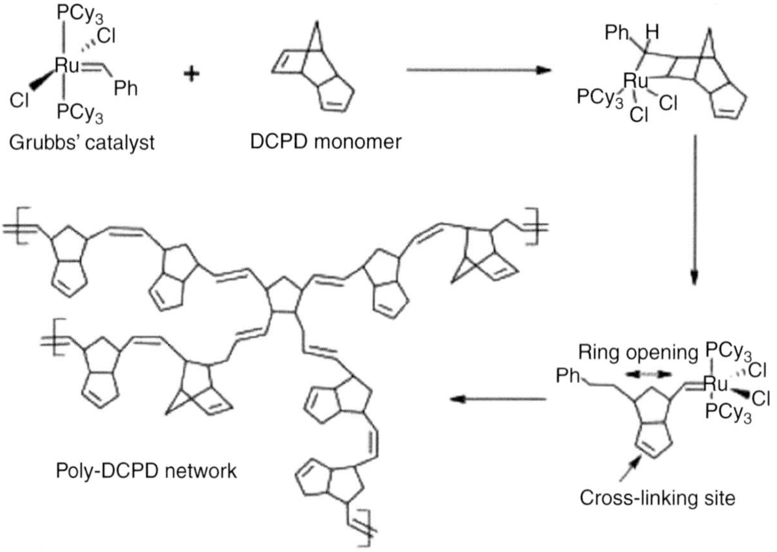

Fig. 1.3 Ring opening metathesis polymerization of DCPD. (Adapted from [19].)

Table 1.1 Important factors for developing microencapsule-based self-healing materials.

Parameters	Influencing factors
Microcapsule	Healing agent must be inert to the polymer shell
	Longer self life of the capsules
	Compatibility with the dispersion polymer medium
	Weak shell wall to enhance rupture
	Proximity to catalyst
	Strong interfacial attraction between polymer matrix and capsule shell wall to promote shell rupture
Monomer	Low viscous monomer to flow to the crack upon capillary action
	Less volatility to allow sufficient time for polymerization
Polymerization	Should be fast
	Stress relaxation and no cure induced shrinkage
	Room temperature polymerization
Catalysts	Dissolve in monomer
	No agglomeration with the matrix polymer
Coatings	Incorporation of microcapsules should have very less influence on physicomechanical properties of the matrix
	Coating thickness must be larger than the microcapsule size
	No clustering of catalysts or microcapsules in the matrix polymer
	Less expensive manufacturing process
Healing	Must be faster
	Multiple

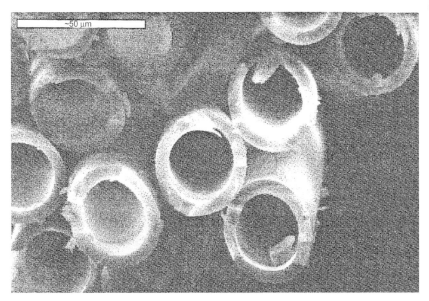

Fig. 1.4 Optical micrographs of hollow glass fibers.
(Reprinted with permission: Dr. J. P. Bond, University of
Bristol, UK.)

Bond and coworkers have employed a biomimetic approach and fabricated composites with bleeding ability. Hollow fibers containing uncured resin or hardener (mixed with UV fluorescent dye for visual inspection) were prepared and plied to achieve a special layered up (0°/90°) structure in the matrix (epoxy resin) in combination with conventional glass fiber/epoxy system. Hollow fiber-based self-healing strategy is shown in Figure 1.5.

They have demonstrated that composite panels prepared using hollow fibers containing repairing agents can restore up to 97% of its initial flexural strength. The release and infiltration of fluorescent dye from fractured hollow fibers into the crack plane was also demonstrated. This approach of self-healing material design offers certain advantages, which are as follows:

- higher volume of healing agent is available to repair damage;
- different activation methods/types of resin can be used;
- visual inspection of the damaged site is feasible;
- hollow fibers can easily be mixed and tailored with the conventional reinforcing fibers.

Besides the above advantages, this approach has the following disadvantages as well:

- fibers must be broken to release the healing agent;
- low-viscosity resin must be used to facilitate fiber infiltration;
- use of hollow glass fibers in carbon fiber-reinforced composites will lead to CTE (coefficient of thermal expansion) mismatch.
- multistep fabrication is required.

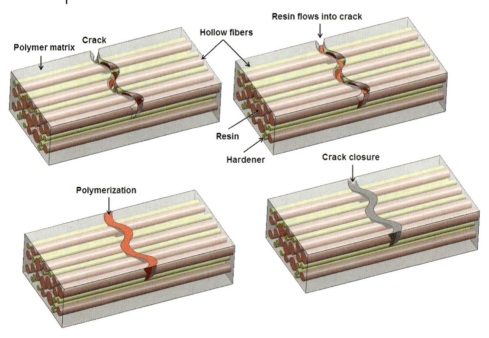

Fig. 1.5 Schematic representation of self-healing concept using hollow fibers.

Recently, Sanada *et al.* have shown the healing of interfacial debonding in fiber-reinforced polymers (FRPs) [42]. They have dispersed microencapsulated healing agent and solid catalyst in the coating layer on the surface of the fibers.

1.3.1.3 Microvascular System

To overcome the difficulty of short supply of a healing agent in microcapsule-based self-healing concept, another approach similar to biological vascular system of many plants and animals was explored by White *et al.* [43, 44]. This approach relies on a centralized network (that is microvascular network) for distribution of healing agents into polymeric systems in a continuous pathway. The fabrication process is complex and it is very difficult to achieve synthetic materials with such networks for practical applications. In this process, organic inks are deposited following a 3D array and the interstitial pores between the printed lines are infiltrated with an epoxy resin. Once the polymer is cured, the fugitive ink is removed leaving behind a 3D microvascular channel with well-defined connectivity. Polymeric systems with microvascular networks were prepared by incorporating chemical catalysts in the polymer used to infiltrate the organic ink scaffold (Figure 1.6). Upon curing the polymer and removing the scaffold, the healing agent is wicked into the microvascular channels. Several researchers reported such fabrication processes and related self-healing capabilities [45–48].

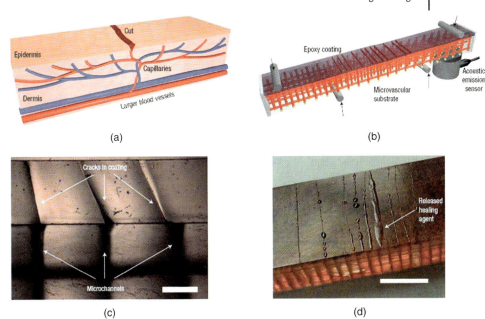

(a)

(b)

(c)

(d)

Fig. 1.6 Schematic showing self-healing materials with 3D microvascular networks. (a) Schematic diagram of a capillary network in the dermis layer of skin with a cut in the epidermis layer. (b) Schematic diagram of the self-healing structure composed of a microvascular substrate and a brittle epoxy coating containing embedded catalyst in a four-point bending configuration monitored with an acoustic-emission sensor. (c) High-magnification cross-sectional image of the coating showing that cracks, which initiate at the surface, propagate toward the microchannel openings at the interface (scale bar = 0.5 mm). (d) Optical image of self-healing structure after cracks are formed in the coating (with 2.5 wt% catalyst), revealing the presence of excess healing fluid on the coating surface (scale bar = 5 mm). [Reprinted with permission from Ref. 44.]

1.3.2
Reversible Cross-links

Cross-linking, which is an irreversible process, of polymeric materials is performed to achieve superior mechanical properties, such as high modulus, solvent resistance, and high fracture strength. However, it adversely affects the refabrication ability of polymers. Moreover, highly cross-linked materials have the disadvantage of brittleness and have the tendency to crack. One approach to bring processability to cross-linked polymers is the introduction of reversible cross-links in polymeric systems [49–51]. In addition to refabrication and recyclability, reversible cross-links also exhibit self-healing properties. However, reversible cross-linked system does not show self-repairing ability by its own. An external trigger such as thermal, photo, or chemical activation is needed to achieve reversibility, and thereby the self-healing ability. Thus, these systems show nonautonomic healing phenomenon. In the following sections, different approaches that are considered to bring reversibility in cross-linked polymeric materials are discussed.

1.3.2.1 Diels–Alder (DA) and Retro-DA Reactions

Major classes of thermally reversible polymers are made using Diels–Alder (DA) reactions. Examples of this category include cross-linking of furanic polymers with maleimide or polymers containing maleimide pendants at low temperature. Retro-DA reaction occurs at elevated temperatures to debond the chemical linkages of formed networks and to reverse the cross-linking process [52]. DA reactions (4 + 2 cycloadditions) are the most studied thermally controlled covalent bond formation. Though there are several reports available on reversible reactions, Wudl and coworkers were the first to implement this strategy to design thermally remendable polymers [53, 54]. The first polymer (3M + 4F = polymer (3M4F)) they synthesized (Figure 1.7) showed a strength recovery of 53% [55]. Later, they have reported improved system with mechanical strength recovery of 83%. Since their discovery, several other research groups around the globe have further contributed to this exciting field of research [56–61].

Liu *et al.* have adapted a modified Wudl's approach in their work [57]. They have synthesized multifunctional furan and maleimide compounds using epoxy compounds as precursors (Figure 1.8). These precursors induce advantageous characteristics of epoxy resins, such as solvent and chemical resistance, thermal and electrical properties, and good adhesion in the final cured polymer. Besides that, the modified furan and maleimide monomers become soluble in most common organic solvents such as acetone, methanol, ethanol, and tetrahydrofuran. The use of solvents, such as acetone, with low-boiling temperature is beneficial, as curing of the matrix can be avoided in the solvent removal stage.

Equal amounts of the modified monomers were dissolved in acetone to produce a homogeneous solution. Then, the solvent was removed and the film was heated

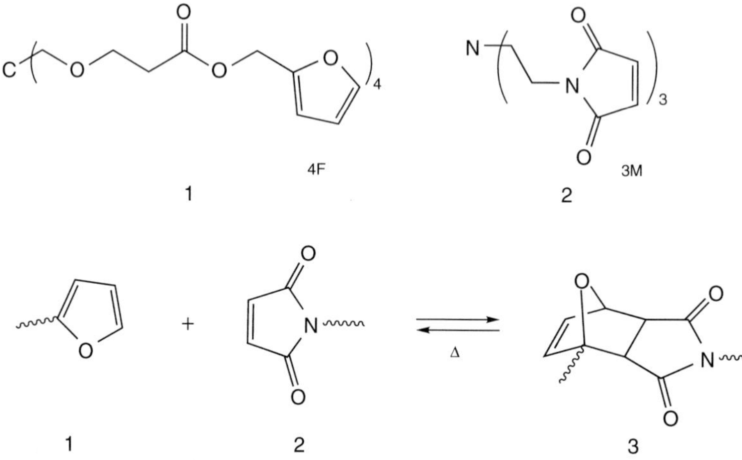

Fig. 1.7 Schematic showing formation of highly cross-linked polymer (3M4F) [polymer 3] using a multi-diene (four furan moieties, 4F) [monomer 1] and multi-dienophile (three maleimide moieties, 3M) [monomer 2] via DA reactions [adapted from Ref. 55].

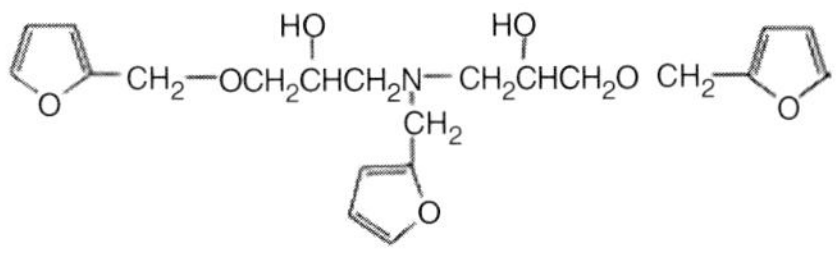

TF

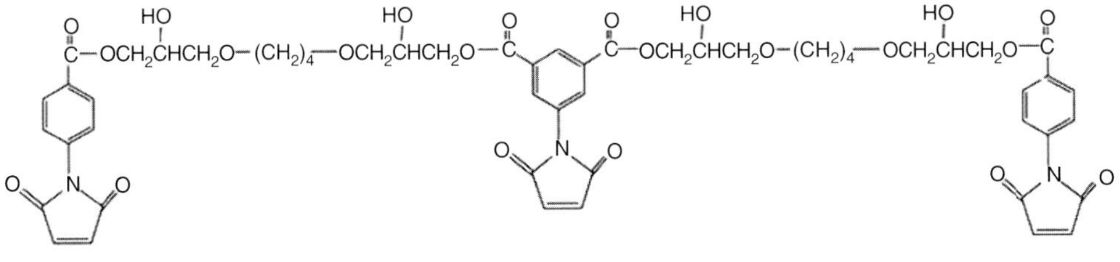

TMI

Fig. 1.8 Chemical structure of functionalized maleimide and furan monomers [adapted from Ref. 57].

in an oven for 12 h at 50 °C. A solid film was produced as the cross-linking between trimaleimide (TMI) and trifuran (TF) takes place via maleimide and furan groups through DA reactions (Figure 1.9). The debonding (retro-DA) occurred upon heating the sample at 170 °C for 30 min.

The self-repairing property of TMI–TF cross-linked material was investigated by morphological analysis using Scanning Electron Microscopic (SEM) techniques (Figure 1.10). The cross-linked material shows a smooth and planar surface as-prepared Figure 1.10a. Figure 1.10b shows a notch made on the surface of the sample by knife-cutting. The cut sample was then thermally treated at 120 °C for 20 min and at 50 °C for 12 h (Figure 1.10c). At higher temperature, debonding (retro-DA) occurred and the polymer chains reformed at this temperature. DA reactions (bonding of polymer chains) take place again at lower temperatures and the cross-linked structure is reformed. A complete repairing was obtained by treating the sample at 50 °C for 24 h (Figure 1.10d).

Later Lu *et al.* have synthesized polymers-based maleimide-containing polyamides and a tri-functional furan compound [58]. The prepared adduct shows good thermoreversibility and gel formation through DA and retro-DA reactions (Figures 1.11 and 1.12).

Recently they have used modified polyamides having various amounts of maleimide and furan pendant groups to obtain self-healing capability using DA and retro-DA reactions [59]. However, the prepared adduct does not show complete repairing of the cracks due to the low mobility of high molecular polyamide chains in bulk.

Recently Chung *et al.* [62] have reported for the first time light-induced crack healing. They have chosen 2 + 2 photochemical cycloaddition of cinnamoyl groups to obtain self-healing properties. Photo-cross-linkable cinnamate monomer, 1,1,1-tris-

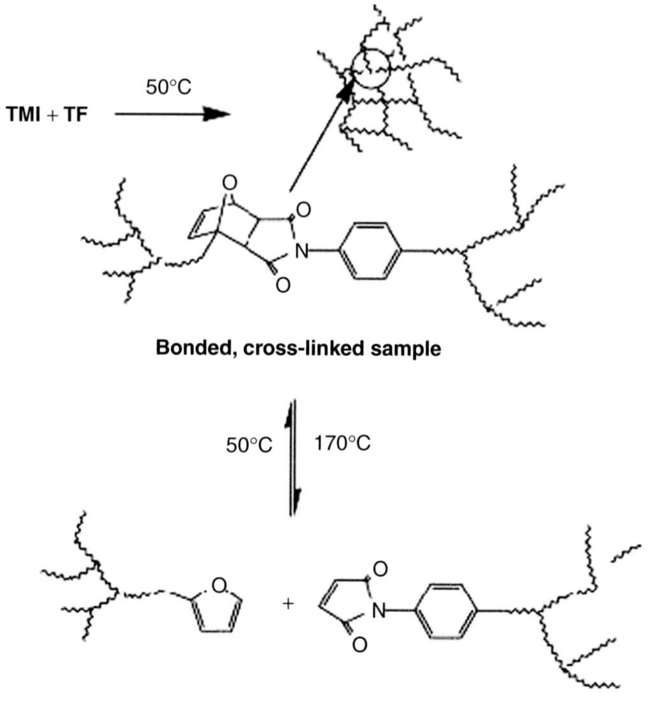

Bonded, cross-linked sample

De-bonded sample

Fig. 1.9 Thermally reversible cross-linking reaction between TMI and TF through DA and retro-DA reactions [Adapted from Ref. 57].

(cinnamoyloxymethyl)ethane (TCE), was used for their study. The photocyclo-addition and recycloaddition of cinnamoyl groups are schematically shown in Figure 1.13.

The authors demonstrated the self-healing capability of the complexes by measuring the flexural strength of cracked and healed samples and the reaction was confirmed by Fourier Transform Infrared (FTIR) spectroscopy. The photochemical healing is very fast and does not require catalysts, additives, or heat treatments.

1.3.2.2 Ionomers

Ionomers are a special class of polymeric materials that contain a hydrocarbon backbone and pendent acid groups, which are neutralized partially or fully to form salts [63–69]. The ion content of ionomeric polymers or ionomers varies over a wide range, but in general it is up to 15 mol%. The methods of synthesis of ionomers can be broadly divided into two main classes: (i) direct synthesis (copolymerization of a low-level functionalized monomer with an olefinic unsaturated monomer) and (ii) post-functionalization of a saturated preformed polymer. The ionic interactions present in ionomers usually involve electrostatic interactions between anions, such as carboxylates and sulfonates, and metal cations from Group 1A, Group 2A, or transitional metal cations. A wide variety of carboxylates, sulfonates or ionomers

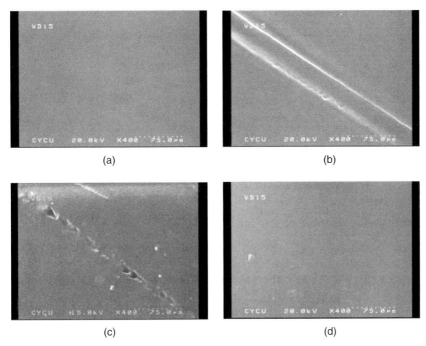

Fig. 1.10 SEM micrographs of (a) cross-linked adducts, (b) knife-cutting sample, and thermally self-repaired sample at 50 °C for (c) 12 h and (d) 24 h [Adapted from Ref. 57].

having both carboxylated and sulfonated groups in the same chain can be found in the literature. The polar ionic groups tend to aggregate as a result of electrostatic interactions despite the opposing tendency of the chain elastic forces. The presence of ionic groups and their interactions produce physical cross-links that are reversible in nature (Figure 1.14).

Introduction of a small amount of ionic group causes dramatic improvement in polymer properties, such as tensile strength, tear resistance, impact strength, and abrasion resistance. As ionomers are not thermosetting materials, they can be processed like thermoplastics. This unique combination of physical properties and processing ease has led this class of polymers to be used in food packaging, membrane separation, roofing materials, automobile parts, golf ball covers, coatings, and so on. Besides the above-mentioned applications, the reversible nature of ionic bonds makes them suitable for designing self-healing polymeric systems [70–72]. A detailed discussion on ionomer morphology and its potential as a self-healing material can be found in Chapter 3.

1.3.2.3 Supramolecular Polymers
Polymeric properties in traditional polymers are achieved due to the length and entanglement of long chains of monomers, which are held together by covalent bonds. Recently, low molar mass monomers are assembled together by reversible noncovalent interactions to obtain polymer-like rheological or mechanical

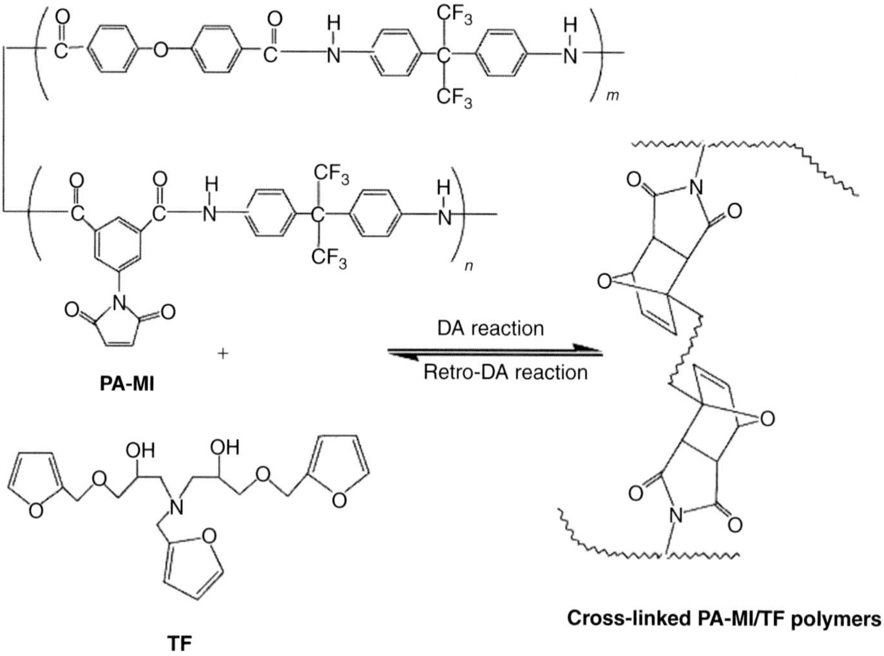

PA-MI

+

TF

DA reaction

Retro-DA reaction

Cross-linked PA-MI/TF polymers

Fig. 1.11 Preparation of thermally reversible polyamides. (Reprinted with permission from [58].)

properties [73–76]. As noncovalent interactions can be reversibly broken and can be under thermodynamic equilibrium, this special class of macromolecular materials, that is, the so-called supramolecular polymers show additional features compared to usual polymers. These features include switchable environment-dependent properties, improved processing, and self-healing behavior. In general, supramolecular polymers can be divided broadly into two categories, which are main- and side-chain types. Although noncovalent interactions hold the backbone of the main-chain supramolecular polymers, it is used to either change or functionalize conventional covalent polymers in case of side-chain supramolecular polymers. Some examples of both classes of supramolecular polymers are shown in Figure 1.15.

Different types of assembly forces, such as metal–ligand interactions, $\pi-\pi$ interactions, hydrophobic, electrostatic interactions, and hydrogen bonding are used to design supramolecular polymers. Hydrogen bonding is the most popular route of achieving supramolecular polymers. The main challenge in this approach is to find the right balance between the association constant and a reversible system. The higher the association constant, the lesser is the reversible interaction. In contrast, the lower the association constant, the better the reversibility, that is, smaller assemblies and poor mechanical properties.

Meijer and coworkers were the first to assemble ureidopyrimidone (Upy) monomers by using quadruple hydrogen bonding noncovalent interactions with

(a)

(b)

(c)

(d)

(e)

(f)

(g)

Fig. 1.12 Photographs showing thermally reversible cross-linking behavior of PA-MI/TF polymers (PA-MI-1/TF polymers have lowest cross-link density and PA-MI-10/TF polymers have highest cross-link density). Polymer gel of PA-MI-1/TF in N, N-dimethylacetamide (DMAc): (a) 30 °C, (b), 160 °C and cross-linked PA-MI-1/TFin DMAc: (c) 30 °C, 5 h, insoluble and (d) 120 °C, 2 h, soluble. Cross-linked PA-MI-10/TF polymer in DMAc: (e) 30 °C, 5 h insoluble, (f) 120 °C, 5 h partially soluble, and (g) 160 °C, 5 h, soluble. (Reprinted with permission from [58].)

high degree of polymerization [77, 78]. The resulting material display mechanical properties similar to traditional polymers. This discovery of using weak reversible hydrogen bonding interactions to produce supramolecular assemblies with high association constant and having polymeric properties makes this field an exciting area for materials research. The Upy compounds are cheap and can be incorporated into other polymeric systems to improve processability or other functionalities. This hydrogen bonded unit is further utilized in the chain extension of telechelic

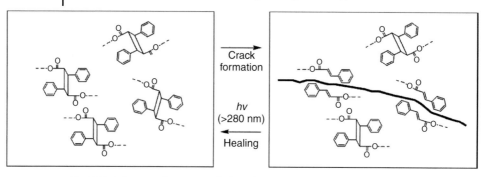

Fig. 1.13 Schematic illustration of photochemical self-healing concept [reprinted with permission from Ref. 62].

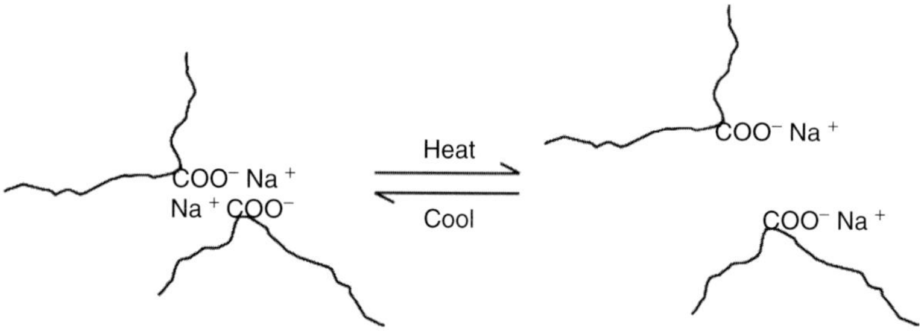

Fig. 1.14 Schematic showing reversible ionic interactions.

polysiloxanes, polyethers, polyesters, and so on. On the basis of the above discovery, a spin-off company from the Technical University of Eindhoven, SupraPolix BV, has already started exploring this field commercially [79]. Hybrid systems were also developed by using supramolecular monomers with traditional polymers. Cordier and colleagues have recently published a very interesting piece of research that brings together supramolecular chemistry and polymer physics to develop self-healing rubbers [80]. They have used fatty diacids and triacids from renewable resources and used two-step synthetic routes to produce self-healing rubbers. In the first step, acid groups were condensed with excess of diethylene triamine, and in the second step, the condensed acid groups were made to react with urea. The resulting material shows rubber-like characteristics and self-healing capability. The prepared material can be repaired by simply bringing the two cut ends together at room temperature without the need of external heat. However, if the broken parts are kept for a longer period (Figure 1.16), they need to be hold together for longer period for self-mending.

Besides hydrogen bonding, metal–ligand supramolecular interactions are also being explored to design supramolecular polymers [81, 82, 83]. Metal complexes offered certain advantages due to its optical and photophysical properties. Moreover, its reversibility can be tuned by using different metal ions. Though bi-pyridine complexes are well known, it is the terpyridine-based metal–ligand complexes that are gaining increased attention as a new type of functional materials (Figure 1.17).

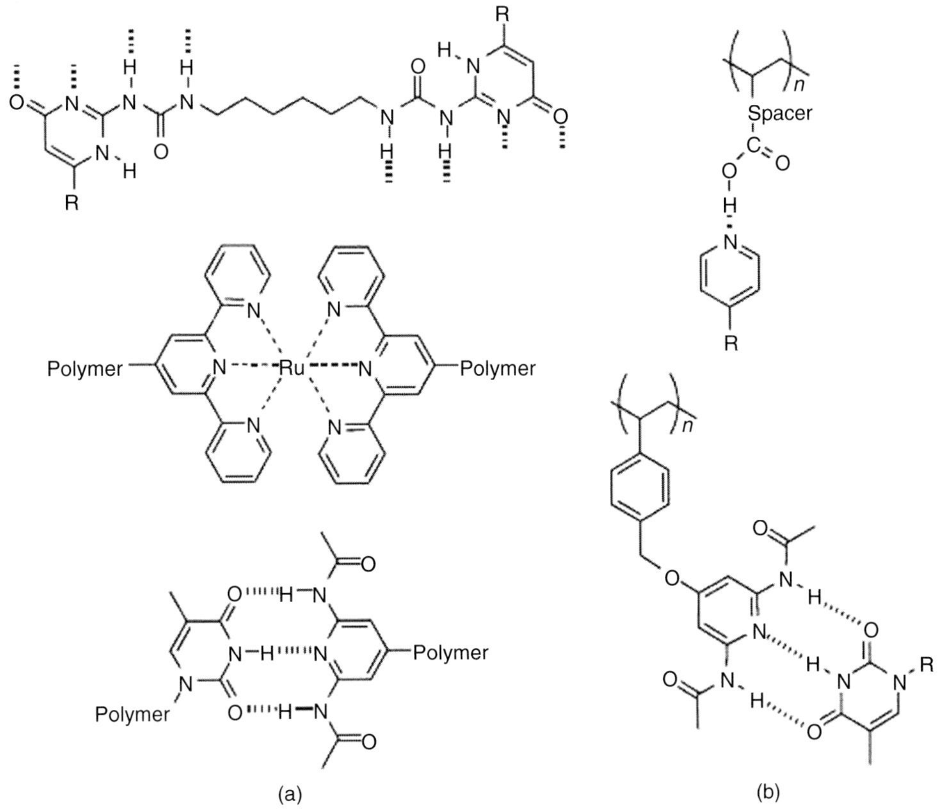

Fig. 1.15 Examples of supramolecular polymers from the literature: (a) main-chain supramolecular polymers and (b) side-chain supramolecular polymers.

These ligands can be introduced into polymeric systems by several ways, such as copolymerization of functionalized monomer, functionalization of end or side groups of preformed polymers, or by using functionalized initiators and/or end cappers in living or controlled polymerizations.

1.3.3
Miscellaneous Technologies

Technologies other than the most important self-healing approaches described above are available in the literature. These emerging technologies are discussed in the following sections of this review.

1.3.3.1 Electrohydrodynamics
In this approach, the blood clotting process was mimicked via colloidal particle aggregation at the defected site. Prof. Ilhan Aksay and his collaborators have used

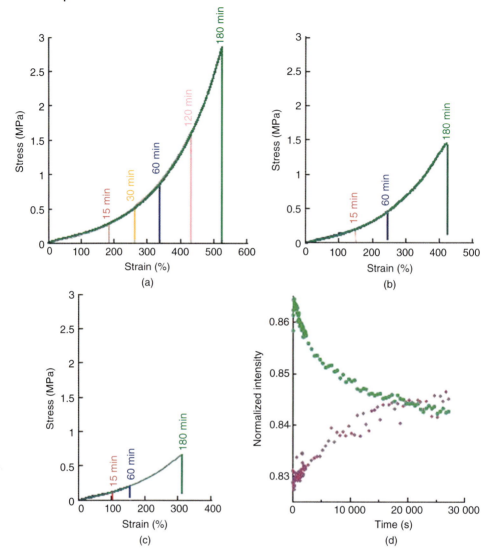

Fig. 1.16 Self-mending at room temperature. (a) Cut parts are brought into contact at 20 °C immediately after being cut (waiting time <5 min). Curves represent stress–strain behavior measured for convenience at 40 °C after different healing times. (b) Stress–strain behavior of mended samples at 40 °C; mending was performed at 20 °C after keeping broken samples apart for 6 h. (c) As in (b) but cut samples were kept apart for 18 h and then mended at 20 °C. Colored vertical lines in (a–c) correspond to elongation at breaking for given healing times (for all healing times, stress–strain curves superpose almost exactly and show elongation only at break changes). (d) Time-dependent infrared experiments. The sample was heated at 125 °C for 10 min and then quenched to 25 °C. Infrared absorption spectra evolutions were recorded. The intensity at 1524 cm^{-1}, characteristic of free N–H bending motions (green), decreases, whereas the intensity at 1561 cm^{-1}, characteristic of associated N–H bending motions (purple), increases. These data confirm the long lifetime of open hydrogen bonds when they are created in excess [reprinted with permission from Ref. 80].

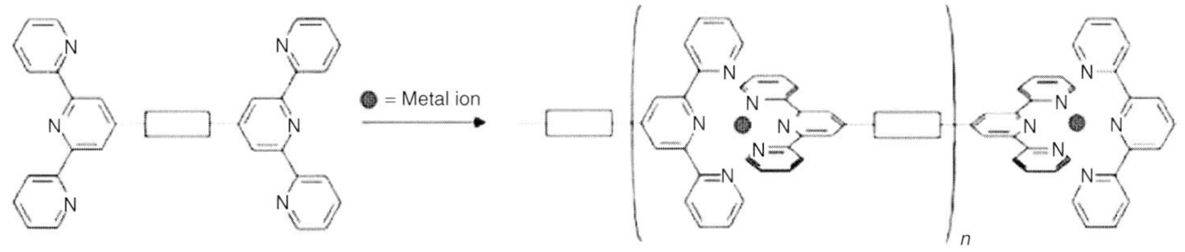

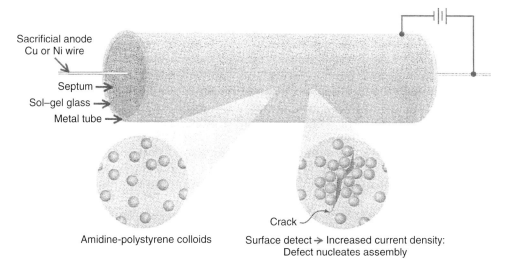

Fig. 1.17 Polymeric bis-terpyridine-metal complex (charge and anions omitted) [adapted from Ref. 81].

Sacrificial anode
Cu or Ni wire

Septum →
Sol–gel glass →
Metal tube →

Amidine-polystyrene colloids

Crack

Surface detect → Increased current density:
Defect nucleates assembly

Fig. 1.18 Schematic showing electrohydrodynamic aggregation of particles [reprinted with permission from Ref. 83].

the principle of electrohydrodynamics (EHD) flow to design self-healing materials [84]. They have used suspension of colloidal particles, which is enclosed between the walls of a double-walled metallic cylinder (Figure 1.18). These walls are coated with a conductive layer followed by a ceramic insulating layer. A concentric metal wire is used to apply electric field to this system.

When damage occurs in the insulating layer, the current density at the damaged site is increased causing an agglomeration of the colloidal particles at the defected site through EHD flow. The aggregation of particles is not sufficient to heal the defects as the voids between colloidal particles prevent formation of a dense surface. The author suggests the use of polymeric colloidal particles or a sacrificial anode for simultaneous electrodeposition of metal at the defect site to achieve better healing efficiency. Thomas *et al.* have reported a concept of self-healing structural composites with electromagnetic functionality [85]. The self-healing is achieved through contributions of all components such as thermoreversible polymers, reinforcing fibers, and electromagnetic wires. The incorporated wires serve as both electrical and thermal conductor and distribute heat uniformly. The added fibers also contribute to the healing mechanism. For example, when fibers having negative CTE is used to fill the core of the braid or fill in the weave of laminate, it will contract upon heating. This forces the matrix to compress and close the defect as the cracked polymer matrix (having positive CTE) expands upon contraction of the reinforcing fibers.

1.3.3.2 Conductivity

Polymeric materials are insulative in nature. By imparting conductivity into polymeric systems these materials can be made suitable for electronic applications. The tunable conductivities in polymeric materials can offer information on the structural integrity through electronic feedback that might give an insight to the most challenging task of detecting and quantifying microcracks. Thus, materials having conductivity as well as self-healing capability might be advantageous especially in deep sea or space applications. The conductivity, on the other hand, can also be used for inducing self-healing properties in polymeric systems. Williams *et al.* have exploited organometallic polymers based on *N*-heterocyclic carbenes and transition metals to design electrically conductive self-healing materials [86]. These polymers exhibit structurally dynamic characteristics in the solid state and have good processability characteristics. The electrical conductivity of the developed reversible systems is of 10^{-3} S cm^{-1}. Their approach of conductive self-healing material design is schematically shown in Figure 1.19. When a microcrack is formed in a system, it decreases the number of electron percolation pathways and thereby an increase in electrical resistance.

If an electrical source is connected, this drop in conductivity can be triggered to increase the applied electric field. Thus, if the rise in resistance is due to microcracking, then this voltage bias can generate localized heat at the microcrack, which can force the system back to its original state, that is, low resistance/high current situation. The organometallic polymeric systems based on *N*-heterocyclic carbenes and metals can be reversibly formed, which meet the conductivity

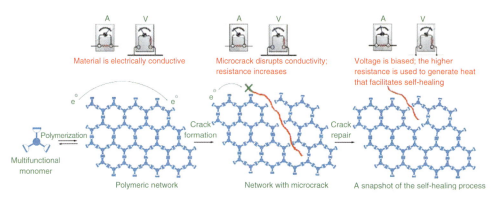

Fig. 1.19 Schematic showing conductive self-healing materials (A = amperes = volt) [reprinted with permission from Ref. 85].

requirements that make them suitable for self-healing applications. The author recommended that incorporation of bulky N-alkyl moieties into carbenes may reduce the viscosity upon depolymerization, which will enhance its flow into the cracks. Moreover, higher conductivities ($\sim$1 S cm^{-1}) should be achieved to have practical self-healing applications. Thostenson and his colleagues have successfully incorporated multiwalled nanotubes (MWNTs) in glass fiber–epoxy composites [87]. It was shown that a very low concentration of carbon nanotubes (0.1 wt%) is sufficient to achieve the percolation threshold in the prepared composites. The MWNT networks in the epoxy composite matrix can also accurately detect the onset, nature, and progression of damage. This property may be useful to have broad applications, including assessing self-healing strategies.

1.3.3.3 Shape Memory Effect

Certain strongly ordered intermetallic systems show the well-known shape memory effect, in which plastic deformation imparted to the low-temperature martensite phase can be reversed almost completely during transformation to the high-temperature austenite phase [88, 89]. These shape memory alloys (SMAs) can be used as self-healing materials. For example, SMAs such as Nitinol (nickel–titanium) exhibits the self-healing effect when heated [90]. If they are permanently deformed and heated above certain temperatures, they will return to their original shape (Figure 1.20). The transformation temperatures at which the alloys have their highest yield strength can be tuned between 100 and $-100\,^\circ$C by

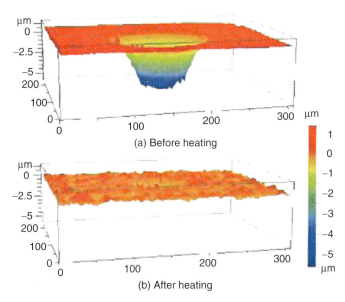

Fig. 1.20 Representative three-dimensional profiles of a spherical indent at load of 15 N: (a) fresh indent and (b) after heating above the austenite finish temperature [reprinted with permission from Ref. 90].

controlling the material properties. A detailed discussion on the use of SMAs as self-healing materials can be found in Chapter 8.

1.3.3.4 Nanoparticle Migrations

Balazs and coworkers have demonstrated that nanoparticles in a polymer fluid can segregate into cracks due to the polymer-induced depletion attraction between the particles and the surface [91–95]. The obtained morphology from the molecular dynamics simulations was used in a lattice spring model to determine the self-healing efficiency. The obtained model predicts restoration of mechanical properties up to 75–100%. Self-healing materials based on the above approach are yet to be demonstrated. Incorporation of nanoparticles into polymeric systems has twofold benefits: it increases the mechanical strength of the system and also segregates to the crack surface. Carbon nanotube is a potential candidate for developing self-healing materials based on this approach due to its superior mechanical properties compared to other particles.

1.3.3.5 Co-deposition

Electrolytic co-deposition can also be employed to design self-healing anticorrosive coatings (Figure 1.21). Microcapsules containing corrosion inhibitors can be added to composite plating coatings by this method [96–98]. Either liquid corrosion inhibitors or mesoporous nanoparticles containing absorbed corrosion inhibitors can be used as the core material to synthesize micro- or nanocapsules [99]. These capsules can be later deposited with metallic ions such as Zn^{+2} and Cu^{+2} to form composite metallic coatings. Upon crack formation in the composite layer, the capsule can release its contents to heal the crack.

Besides the above-mentioned self-healing strategies, many other approaches are expected to come in the near future for the development of self-healing materials. In this context, it is also important to note the different ways to evaluate quantitatively

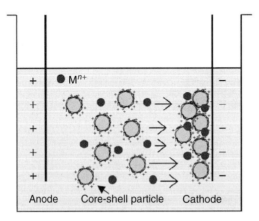

Fig. 1.21 Schematics showing electrolytic co-deposition of microcapsules (or mesoporous nanoparticles containing corrosion inhibitors) with metal ions.

the healing efficiency as different systems or authors use several methodologies to evaluate their systems. A summary of these quantitative routes can be found in the review of Kessler [5].

1.4
Applications

Product commercialization in industries is usually based on the following major milestones: idea generation (preliminary level) → laboratory implementation (product level) → pilot line up scaling (process level) → industrial applications (marketing level). Currently, self-healing materials development is either in the preliminary or product level, and so these materials are yet to be available for many applications. Applications of self-healing materials are expected almost entirely in all industries in future. The very few applications being developed to date are mainly in the automotive, aerospace, and building industries. For example, Nissan Motor Co. Ltd has commercialized world's first self-healing clear coat for car surfaces. The trade name of this product is "Scratch Guard Coat" [7]. According to the company, this hydrophobic paint repairs scratches (arising from car washings, off-road driving, or fingernails) on coated car surfaces and is effective for a period of three years. This newly developed paint contains high elastic resins that prevent scratches reaching the inner layers of a painted car surface. Depending on the depth of the scratch and the temperature in the surrounding environment, the entire recovery occurs between 1 and 7 days. Another example in this category is the two component polyurethane clear coats from Bayer MaterialScience [100]. The trade names of the raw materials used to formulate this coating are Desmodur and Desmophen. According to company sources (Figure 1.22), this coating heals small scratches under the influence of heat (sunlight) and the trick employed to design such coatings is based on the use of dense polymer networks with flexible linkages.

For both the above examples the scratch discussed is in the range of few micrometers, which is obviously visible to the naked eye, and therefore the products are suitable for keeping the aesthetics of the coating. Moreover, the above examples also follow similar self-healing mechanisms. Energy required to overcome the resistance of materials to create a scratch is higher in the case of thermosetting polymers (proportional to its plastic and/or elastic response) compared to thermoplastic polymers (viscoelastic response). Formation of a scratch in materials leads transport of materials from the affected zone to its side leaving the groove. In case of thermoplastic polymers, the energy is lost in the process of viscous flow in the absence of residual stress (due to viscoelastic or plastic deformation). Thus most important driving force that helps the reflow of materials from the side to the groove is surface tension. However, for thermosetting polymers, the energy (below it yield's strength) incorporated to create a scratch is stored in the neighborhood of the conduit. When the mechanical stress is removed, the stored energy is relieved and the distorted polymer chains returns leveling the groove. This recovery process is highly dependent on the mobility of the polymer chains

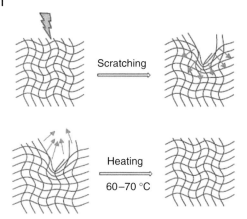

Scratching

Heating

60–70 °C

Fig. 1.22 Schematic showing the reflow effect of self-healing clear coats [adapted from a presentation [100] of Bayer Materials Science].

that is on their glass transition temperature (T_g). However, while scratching, if the mechanical stress also leads to cracking besides scratch formation, the stored energy will be released at the inappropriate time and a partial recovery (plastic residual strain allows some reflow) may be expected as surface tension-driven viscous process will not take place here due to the presence of opposing elastic force in the system. Thus, scratch with fractures is a permanent damage for thermosetting polymers, and therefore a compromise has to be considered between the above two processes for designing self-healing polymer coatings. External trigger can be useful in this case. Thus, polymers with high T_g (less material transport) in combination with high elastic response could be an option for the recovery of small scratches. In case of small fractures, triggering by temperature will enhance the mobility of the polymer chains and surface tension will play an important role for self-healing.

The next industrial segment where applications of self-healing materials are foreseen is the aviation industry. Use of composites in aircrafts has grown significantly in the past years. Hollow fibers reinforced composites are a possible solution to recover cracking or damages. Self-healing polymers have paved its way in space applications.

The construction industry will also find many applications of self-healing materials. For example, self-healing concretes may become a reality soon. Self-healing corrosion resistant coatings could be beneficial for structural metallic components such as steel for achieving long-term service life with reduced maintenance cost. Other areas of applications of self-healing materials are in medical segments. Biocompatible self-healing composite may extend the service life of artificial bone, artificial teeth, and so on. The very recent discovery of self-healing rubber may find applications in the toy industry.

Finally, it could be said that the available technologies to design self-healing materials are not cost effective. This limits the wide use of these materials for different commercial applications. In future, one can expect to see new technologies that will enable the possibility of using self-healing materials in our day-to-day life.

1.5
Concluding Remarks

Microcracking and hidden damages are the initiators for structural failures. On the other hand, high maintenance and repairing costs limit the acceptance of different materials in engineering disciplines. Repairing at remote locations is very difficult. In this context, self-healing materials possess tremendous potential in increasing the longevity of structural materials. Consequently, a large number of academic and industrial research organizations have come forward to explore new concepts in this promising filed. A growing trend of investments is also being seen from both the government and industrial funding agencies. Though this field of innovative product research shows high promises, it has some practical limitations in understanding crack healing kinetics and stability of healing functionality. Thus the main challenge of self-healing material development is autonomic detection of cracks and its subsequent healing.

To be able to develop new innovative solutions based on biomimetic approaches, it is an ultimate need to overcome the difficulties of damage detection and achieving an autonomic-healing phenomenon. Besides autonomic-healing, nonautonomic processes will also find industrial applications. It is sure that this wonderful field of self-healing materials will continue to grow beyond the technologies reviewed here and it will become available for our daily uses.

References

1 Weiner, S. and Wagner, H.D. (1998) *Annual Review of Materials Science*, **28**, 271–98.

2 Zhou, B.L. (1996) *Materials Chemistry and Physics*, **45** (2), 114–19.

3 Fratzl, P. and Weinkamer, R. (2007) in *Self Healing Materials. An Alternative Approach to 20 Centuries of Materials Science* (ed S. vanderZwaag), Springer, pp. 323–35.

4 Vermolen, F.J., van Rossum, W.G., Javierre, E. and Adam, J.A. (2007) *Self Healing Materials. An Alternative Approach to 20 Centuries of Materials Science* (ed S. vanderZwaag), Springer, pp. 337–63.

5 Kessler, M.R. (2007) *Proceedings of the Institution of Mechanical Engineers Part G-Journal of Aerospace Engineering*, **221**, 479–95.

6 Wool, R.P. (2008) *Soft Matter*, **4**, 400–18.

7 http://www.nissan-global.com/EN/ TECHNOLOGY/INTRODUCTION/ DETAILS/SGC/index.html. (Access year 2008).

8 Thies, C. (1987) Microencapsulation, in *Encyclopedia of Polymer Science and Engineering*, Vol. 9, John Wiley & Sons, Inc, New York, pp. 724–45.

9 Benita, S. (1996) *Microencapsulation: Methods and Industrial Applications*, Marcel Dekker, New York.

10 Arshady, R. (1999) *Microspheres, Microcapsules and Liposomes*, Citrus Books, London.

11 Ghosh, S.K. (**2006**) *Functional Coatings by Polymer Microencapsulation*, Wiley-VCH Verlag GmbH, Germany.

12 Dry, C. (**1996**) *Composite Structures*, **35**, 263–69.

13 Jung, D., Hegeman, A., Sottos, N.R., Geubelle, P.H. and White, S.R. (**1997**) *The American Society for Mechanical Engineers Materials Division*, **80**, 265–75.

14 White, S.R., Sottos, N.R., Geubelle, P.H., Moore, J.S., Kessler, M.R., Sriram, S.R., Brown, E.N. and Viswanathan, S. (**2001**) *Nature*, **409**, 794–97.

15 Jones, A.S., Rule, J.D., Moore, J.S., White, S.R. and Sottos, N.R. (**2006**) *Chemistry of Materials*, **18**, 1312–17.

16 Brown, E.N., Kessler, M.R., Sottos, N.R. and White, S.R. (**2003**) *Journal of Microencapsulation*, **20** (6), 719–30.

17 Shansky, E. (**2006**) *Synthesis and Characterization of Microcapsules for Self-healing Materials*, Department of Chemistry, Indiana University, Bllomington.

18 Tillner, S. and Mock, U. (**2007**) *Farbe Und Lack*, **10**, 35–42.

19 Larin, G.E., Bernklau, N., Kessler, M.R. and DiCesare, J.C. (**2006**) *Polymer Engineering and Science*, **46**, 1804–11.

20 Rule, J.D. and Moore, J.S. (**2002**) *Macromolecules*, **35**, 7878–82.

21 Cho, S.H., Andersson, H.M., White, S.R., Sottos, N.R. and Braun, P.V. (**2006**) *Advanced Materials*, **18**, 997–1000.

22 Keller, M.W., White, S.R. and Sottos, N.R. (**2006**) An elastomeric self-healing material, in *Proceedings of the 2006 SEM Annual Conference and Exposition on Experimental and Applied Mechanics*, Society for Experimental Mechanics Vol. 1, pp. 379–82.

23 Rule, J.D., Sottos, N.R. and White, S.R. (**2007**) *Polymer*, **48**, 3520–29.

24 *http://www.sciencedaily.com/releases/2007/11/071127105523.htm.* (Access year 2008).

25 Kessler, M.R. and White, S.R. (**2001**) *Composites Part A*, **32**, 683–99.

26 Kessler, M.R., Sottos, N.R. and White, S.R. (**2003**) *Composites Part A*, **34**, 743–53.

27 Brown, E.N. (**2003**) Fracture and fatigue of a self healing polymer composite materials, Ph.D Thesis, University of Illinois at Urbana-Champaign, Urbana-Champaign.

28 Brown, E.N., White, S.R. and Sottos, N.R. (**2004**) *Journal of Materials Science*, **39**, 1703–10.

29 Brown, E.N., White, S.R. and Sottos, N.R. (**2005**) *Composite Science and Technology*, **65**, 2466–80.

30 Rule, J.D., Brown, E.N., Sottos, N.R., White, S.R. and Moore, J.S. (**2005**) *Advanced Materials*, **17** (2), 205–8.

31 Yin, T., Rong, M.Z., Zhang, M.Q. and Yang, G.C. (**2007**) *Composites Science and Technology*, **67**, 201–12.

32 Dry, C.M. (**1995**) *Proceedings of SPIE - International Society for Optical Engineering*, **2444**, 410–13.

33 Dry, C.M. and McMillan, W. (**1996**) *Proceedings of SPIE - International Society for Optical Engineering*, **2718**, 448–51.

34 Motuku, M., Vaidya, U.K. and Janowski, G.M. (**1999**) *Smart Materials and Structures*, **8**, 623–38.

35 Bleay, S.M., Loader, C.B., Hawyes, V.J., Humberstone, L. and Curtis, V. (**2001**) *Composites A, Applied Science Manufacturing*, **32**, 1767–76.

36 Hucker, M., Bond, I., Foreman, A. and Hudd, J. (**1999**) *Advanced Composites Letters*, **8** (4), 181–89.

37 Trask, R.S. and Bond, I.P. (**2006**) *Smart Materials and Structures*, **15**, 704–10.

38 Pang, J.W.C. and Bond, I.P. (**2005**) *Composites Part A-Applied Science and Manufacturing*, **36**, 183–88.

39 Pang, J.W.C. and Bond, I.P. (**2005**) *Composites Science and Technology*, **65**, 1791–99.

40 Williams, H.R., Trask, R.S. and Bond, I.P. (**2006**) *"Vascular Self-Healing Composite Sandwich Structures"*, 15th US National Congress of Theoretical and Applied Mechanics. Boulder, CO, 25–31 June.

41 Williams, G.J., Trask, R.S. and Bond, I.P. (**2007**) *Composites A*, **38** (6), 1525–32.

42 Sanada, K., Yasuda, I. and Shindo, Y. (**2006**) *Plastics, Rubber and Composites*, **35** (2), 67–71.

43 Therriault, D., White, S.R. and Lewis, J.A. (**2003**) *Nature Materials*, **2**, 265–271.

44 Toohey, K.S., Sottos, N.R., Lewis, J.A., Moore, J.S. and White, S.R. (**2007**) *Nature Materials*, **6**, 581–85.

45 Therriault, D., Shepherd, R.F., White, S.R. and Lewis, J.A. (**2005**) *Adv. Mater.*, **17**, 395–399.

46 Kim, S., Lorente, S. and Bejan, A. (**2006**) *J. Appl. Phys.*, **100**, (6), 063525(1–8).

47 Toohey, K.S., White, S.R. and Sottos, N.R. (**2005**) in *Proceedings, of the Society for Experimental Mechanics (SEM) Annual Conference and Exposition on Experimental and Applied Mechanics*, 241–244.

48 Lewis, J.A. and Gratson, G.M. (**2004**) *Mater. Today*, **7**, 32–39.

49 Adhikari, B., De, D. and Maiti, S. (**2000**) *Progress in Polymer Science*, **25** (7), 909–48.

50 Bergman, S.D. and Wudl, F. (**2008**) *Journal of Materials Chemistry*, **18**, 41–62.

51 Bergman, S.D. and Wudl, F. (**2007**) van der Zwaag, S. (ed.) *Self-healing materials. An Alternative Approach to 20 Centuries of Materials Science*, Springer, pp. 45–68.

52 Chen, X., Dam, A., Ono, K., Mal, A., Shen, H., Nutt, S.R., Sheran, K. and Wudl, F. (**2002**) *Science*, **295** (5560), 1698–701.

53 Chen, X., Wudl, F., Mal, A.K., Shen, H. and Nutt, S.R. (**2003**) *Macromolecules*, **36**, 1802–7.

54 Chen, X. (**2003**) Novel polymers with thermally controlled covalent cross-linking, Ph. D Thesis, University of California, Los Angeles.

55 Chen, X., Dam, M.A., Ono, K., Mal, A., Shen, H., Nutt, S.R., Sheran, K. and Wudl, F. (**2002**) *Science*, **295** (5560), 1698–702.

56 Liu, Y.-L. and Wang, Y.-H. (**2004**) *Journal of Polymer Science. Part A-1, Polymer Chemistry*, **42**, 3178–88.

57 Liu, Y.-L. and Hsieh, C.-Y. (**2006**) *Journal of Polymer Science. Part A-1, Polymer Chemistry*, **44**, 905–13.

58 Liu, Y.-L., Hsieh, C.-Y. and Chen, Y.-W. (**2006**) *Polymer*, **47**, 2581–86.

59 Liu, Y.-L. and Chen, Yi.-W. (**2007**) *Macromolecular Chemistry and Physics*, **208**, 224–32.

60 Chang, J.Y., Do, S.K. and Han, M.H. (**2001**) *Polymer*, **42**, 7589–94.

61 Scott, T.F., Schneider, A.D., Cook, W.D. and Bowman, C.N. (**2005**) *Science*, **308** (5728), 1615–18.

62 Chung, C.-M., Roh, Y.-S., Cho, S.-Y. and Kim, J.-G. (**2004**) *Chemistry of Materials*, **16**, 3982–84.

63 Eisenberg, A. and Kim, J.S. (eds) (**1998**) *Introduction to Ionomers*, John Wiley & Sons, New York.

64 Eisenberg, A. and King, M. (**1977**) *Ion Containing Polymers*, Academic Press, New York.

65 Holliday, L. (ed) (**1975**) *Ionic Polymers*, Applied Science Publishers, London.

66 Eisenberg, A. (ed) (**1980**) *Ions in Polymers*, American Chemical Society, Washington, DC.

67 Schlick, S. (ed) (**1996**) *'Ionomers', Characterizations, Theory and Applications*, CRC Press, Boca Raton.

68 Tant, M.R., Mauritz, K.A. and Wilkes G.L. (eds) (**1997**) *Ionomers: Synthesis, Structure, Properties and Applications*, Chapman & Hall, London.

69 Ghosh, S.K. (**1999**) Studies on elastomeric ionomers based on EPDM terpolymer and SEBS block copolymer, Ph. D Thesis, December I.I.T, Kharagpur, India.

70 Fall, R. (**2001**) Puncture reversal of ethylene ionomers—mechanistic studies, Master Thesis, Virginia Polytechnic Institute and State University, Blacksburg, Virginia.

71 Kalista, S.J. Jr. (**2003**) Self-healing of thermoplastic poly(Ethylene-co-Methacrylic Acid) copolymers following projectile puncture,

Master Thesis, Virginia Polytechnic Institute and State University, Blacksburg, Virginia.

72 Sun, C.X., van der Mee, M.A.J., Goossens, J.G.P. and van Duin, M. (2006) *Macromolecules*, 39, 3441–49.

73 Bouteiller, L. (2007) *Advances In Polymer Science*, 207, 79–112.

74 Lehn, J.M. (2002) *Polymer International*, 51, 825–39.

75 Weck, M. (2007) *Polymer International*, 56, 453–60.

76 Shimizu, L.S. (2007) *Polymer International*, 56, 444–52.

77 Beijer, F.H., Sijbesma, R.P., Kooijman, H., Spek, A.L. and Meijer, E.W. (1998) *Journal of the American Chemical Society*, 120, 6761.

78 Söntjens, S.H.M., Sijbesma, R.P., van Genderen, M.H.P. and Meijer, E.W. (2001) *Macromolecules*, 34, 3815.

79 *http://www.suprapolix.com/*. (Access year 2008).

80 Cordier, P., Tournilhac, F., Soulie'-Ziakovic, C. and Leibler, L. (2008) *Nature*, 451, 977–80.

81 Andres, P.R. and Schubert, U.S. (2004) *Advanced Materials*, 16 (13), 1043–68.

82 Hoogenboom, R., Winter, A., Marin, V., Hofmeier, H. and Schubert, U.S. (2007) *American Chemical Society Proceedings*.

83 Ristenpart, W.D., Alesay, I.A. and Saville, D.A. (2007) *Langmuir*, 23 (7), 4071–4080.

84 Aksay, I. (2000) Princton University.

85 Plaisted, T.A., Amirkhizi, A.V., Arbelaez, D., Nemat-Nasser, S.C. and Nemat-Nasser, S. (2003) *Proceedings of SPIE*, 5054, 372–81.

86 Williams, K.A., Boydston, A.J. and Bielawski, C.W. (2007) *Journal of the Royal Society of Interface*, 4, 359–62.

87 Thostenson, E.T. and Chou, T.-W. (2006) *Advanced Materials*, 18, 2837–41.

88 Otsuka, K. and Wayman, C.M. (1998) *Shape Memory Materials*, Cambridge University Press, Cambridge, p. 49.

89 Liu, Y., Xie, Z., Humbeeck, J.V. and Delaey, L. (1999) *Materials Science and Engineering*, 679, A273–75.

90 Nia, W., Cheng, Y.-T. and Grummon, D.S. (2002) *Applied Physics Letters*, 80 (18), 3310–12.

91 Lee, J.Y., Boxton, G.A. and Balazs, A.C. (2004) *Journal of Chemical Physics*, 121, 5531–40.

92 Tyagi, S., Lee, J.Y., Boxton, G.A. and Balazs, A.C. (2004) *Macromolecules*, 37, 9160–68.

93 Smith, K.A., Tyagi, S. and Balazs, A.C. (2005) *Macromolecules*, 38, 10138–47.

94 Gupta, S., Zhang, Q., Emrick, T., Balazs, A.C. and Russell, T. (2006) *Nature Materials*, 5, 229–33.

95 Balazs, A.C., Emrick, T. and Russell, T.P. (2006) *Science*, 314, 1107–10.

96 Stempniewicz, M., Rohwerder, M. and Marlow, F. (2007) *Chem Phys Chem*, 8, 188–94.

97 *http://www.imprs-surmat.mpg.de/ IMPSSurMat/impssurmat.html*.

98 Liqun, Z. (2006) in *Functional Coatings by Polymer Microencapsulation*, (ed S.K., Ghosh), Wiley-VCH Verlag GmbH, Weinheim, Germany, pp. 297–342.

99 Schukin, D.G. and Möhwald, H. (2007) *Small*, 3 (6), 926–43.

100 *http://www.research.bayer.com/ edition_16/Self_healing_automotive_ coating.aspx*. (Access year 2008).

2
Self-healing Polymers and Polymer Composites

Ming Qiu Zhang, Min Zhi Rong and Tao Yin

2.1
Introduction and the State of the Art

Owing to the advantages such as light weight, good processibility, and chemical stability in any atmospheric conditions polymers have been widely used in many modern engineering fields. It is worth noting that, however, long-term durability and reliability of polymeric materials serving for structural applications have puzzled scientists and engineers [1]. The exposure to harsh environment would easily lead to the degradations of polymeric components. Comparatively, microcracking is one of the fatal deteriorations generated in service, which would bring about catastrophic failure of the composites, and hence significantly shorten the lifetimes of structures.

Since the damages inside the materials are difficult to be perceived and to be repaired, in particular, the materials need to have the ability of self-healing. In fact, many naturally occurring materials in animals and plants are themselves self-healing materials [2]. Accordingly, efforts have been made to mimic the natural healing in living bodies and to integrate the bioinspired self-healing capability into polymers and polymer composites. The progress in this aspect has opened an era of new intelligent materials.

Polymers can be separated into two different groups depending on their behavior when heated, that is thermoplastics and thermosets. For thermoplastic polymers, however, self-healing of microcracks has not yet been successful. The approaches applicable to thermosetting polymers, which will be discussed hereinafter in detail, are not suitable for thermoplastics due to the limitation of the material's nature.

Nevertheless, some pilot studies have been conducted in this direction. With the assistance of artificial measures (such as using solvent and heating), for example, cracks can be healed (or welded). Jud and Kaush [3] tested crack healing behavior in a series of poly(methyl methacrylate) (PMMA) samples of different molecular weights and degrees of copolymerization. They induced crack healing by heating the samples above the glass transition temperature under slight pressure. It was found that full resistance was regained during short-term loading experiments. The

Self-healing Materials: Fundamentals, Design Strategies, and Applications. Edited by Swapan Kumar Ghosh
Copyright © 2009 WILEY-VCH Verlag GmbH & Co. KGaA, Weinheim
ISBN: 978-3-527-31829-2

establishment of mechanical strength should result from interdiffusion of chains and formation of entanglements for the glassy polymer [4].

Later, a widely accepted theory of crack healing was proposed by Wool and O'Connor [5], who explained healing in terms of five stages: surface rearrangement, surface approach, wetting, diffusion, and randomization. Accordingly, Kim and Wool [6] proposed a microscopic model for the last two stages on the basis of reptation model that described the longitudinal chain diffusion responsible for crack healing. Wool [7] suggested that the recovery of fracture stress is proportional to $t^{1/4}$ for polybutadiene, where t is the period of heating treatment. Jud et al. [8] also performed rehealing of glassy polymers (PMMA–PMMA, styrene–acrylonitrile copolymer (PSAN)–PSAN, and PMMA–PSAN) at temperatures above the glass transition temperatures, and found that the fracture toughness K_{li} in the interface increased with contact time, t, as $K_{li} \propto t^{1/4}$ as predicted by the diffusion model.

It is worth noting that while craze healing occurs at temperature above and below the glass transition temperature [9], crack healing happens only at or above the glass transition temperature [10]. In order to reduce the effective glass transition temperature of PMMA, Lin et al. [11] and Wang et al. [12] treated PMMA with methanol and ethanol, respectively. They reduced the glass transition temperature to a range of 40–60 °C, and found that there were two distinctive stages for crack healing: the first one corresponding to the progressive healing due to wetting, and the second relating to diffusion enhancement of the quality of healing behavior.

In fact, cracks and strength decay might be caused by structural changes of atoms or molecules, such as chain scission. Therefore, inverse reaction, that is recombination of the broken molecules, should be one of the repairing strategies. Such a method does not focus on cracks healing but on "nanoscopic" deterioration. One example is polycarbonate (PC) synthesized by ester exchange method. The PCs were treated in a steam pressure cabin at 120 °C prior to the repair [13]. As a result, molecular weight of the PCs dropped by about 88–90%. After drying them in a vacuum cabin, the repairing treatment was done in an oven at 130 °C with N_2 atmosphere under reduced pressure. The reduced tensile strength due to the deterioration treatment can, thus, be gradually recovered. The repairing mechanism was considered as follows. First, the carbonate bond was cut by hydrolysis, and so the concentration of the phenoxy end increased after deterioration. Then, the (–OH) end group on the chain was substituted by sodium ion. The –ONa end might attack a carbonate bond at the end of one of the other chains, leading to recombination of these two chains with the elimination of the phenol from PC (Figure 2.1). The repairing reaction was accelerated by weak alkaline medium, such as sodium carbonate. It suggested that two conditions are required for the PC to recombine the polymer chains. One is the chemical structure of the chain end and the other is the catalyst (Na_2CO_3) for acceleration of the reaction.

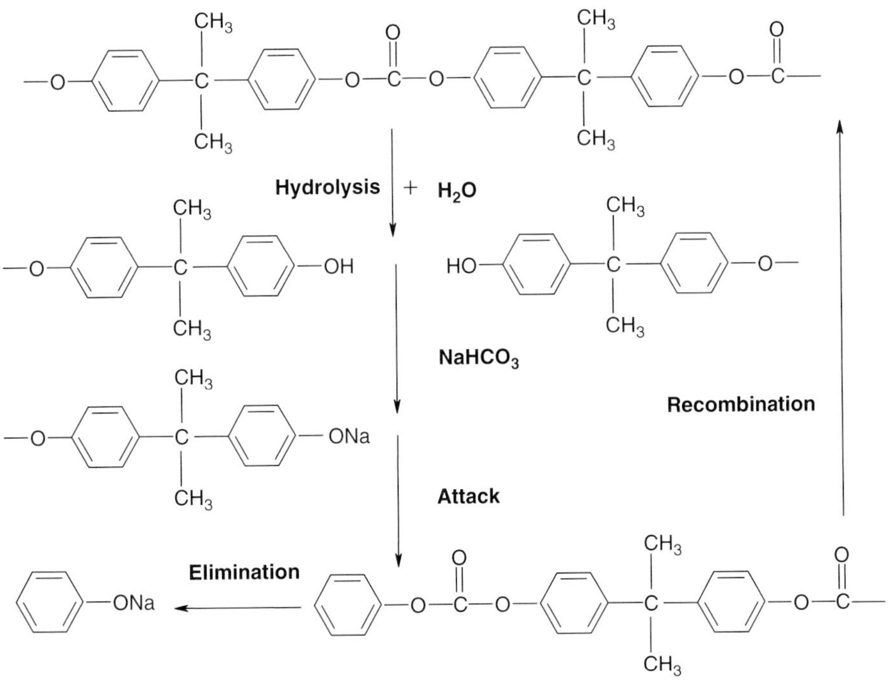

Fig. 2.1 Hydrolysis and recombination reaction of PCs with the catalyst of Na_2CO_3.

Another example is polyphenylene-ether (PPE) in which the repairing agent was regenerated by oxygen [14]. The polymer chain of the PPE was cut by a deterioration factor (such as heat, light, and external mechanical force) to produce a radical at the end of the scission chain. Subsequently, the radical was stabilized by a hydrogen donor. The catalyst existing in the system, Cu(II), reacted with each end of the scission chains to form a complex. Then, the chains were combined by eliminating two protons from the ends, and the Cu(II) is reduced to Cu(I). Afterward, two Cu(I) reacted with an oxygen molecule and were oxidized to Cu(II), and an oxygen ion reacted with two protons to form the water molecule that was lost from the specimen.

The above examples show that PC or PPE might be probably designed as a self-repairing material by means of the reversible reaction. The deterioration is expected to be minimized if the recovery rate is the same as the deterioration rate. However, the systems in these studies are not sufficient for construction of real self-repairing composites because the recovery of the broken molecules needs higher temperature and other rigorous conditions. A much more effective catalyst should be found out, which is able to activate the recombination of degraded oligomers at room temperature.

The achievements in making self-healing thermosetting materials are comparatively more significant. So far, self-healing of thermosetting polymers and their composites fall into two categories: self-healing with and without the aid of healing agent. The works by Chen *et al.* represent the former class [15, 16]. They synthesized highly cross-linked polymeric materials using multifuran and multimaleimide via Diels–Alder reaction. At temperatures above 120 °C, the "intermonomer" linkages disconnect, but then reconnect upon cooling. This process is fully reversible and can be used to restore fractured parts of the polymers.

For healing agent aided self-mending, the agent should be in liquid form at least at the healing temperature. It is generally encapsulated and embedded into the composites' matrix. As soon as the cracks destroy the capsules, the healing agent will be released into the crack planes due to capillary effect and heal the cracks. According to the compositions, the healing agents can be classified as single- and two-component ones. The single-component healants, such as cyanoacrylate [17, 18] and polyvinyl acetate [19], are characterized by low viscosity, wide adaptability, and fast consolidation even at lower temperatures. They are cured under the induction of air, and hence not suitable for healing damages deep in the composites. The two-component system consists of polymerizable resin and hardener. When they meet, polymerization is activated so that the cracked parts can be bonded. Usually, encapsulation of the healing agents is done using fragile-walled containers. The glass pipette tubes were dry filled with cyanoacrylate and epoxy adhesive in the 1:2 ratio [17, 18]. Similar approach was adopted by Motuku *et al.* [20] and Zhao *et al.* [19]. Because the hollow glass capillaries have diameters (on millimeter scale) much larger than those of the reinforcing fibers in composites, they have to act as initiators for composites failure [21]. To minimize the detrimental effect associated with the large diameter fibers, Bleay *et al.* employed hollow glass fiber with an external diameter of 15 μm and an internal diameter of 5 μm [21], but filling of healing agents into such fine tubes becomes rather difficult. Recently, Trask *et al.* [22] considered the placement of self-healing hollow glass fiber layers within both glass fiber/epoxy and carbon fiber/epoxy composite laminates to mitigate damage and restore mechanical strength. The study revealed that after the laminates were subjected to quasi-static impact damage, a significant fraction of flexural strength can be restored by the self-repairing effect of a healing resin stored within hollow fibers. More details of such healing system can be found in Refs 23–26.

On the other hand, Jung *et al.* used polyoxymethyleneurea (PMU)-walled microspheres to store an epoxide monomer to be released into cracks and rebond the cracked faces in a polyester matrix [27]. Solidification of the epoxy resin (i.e. the repair action) was triggered by the excessive amine in the composites. White *et al.* indicated that the method was not feasible as the amine groups did not retain sufficient activity [28]. Zako *et al.* proposed an intelligent material system using 40% volume fraction unmodified epoxy particles to repair microcracks and delamination damage in a glass/epoxy composite laminate [29]. By heating to 120 °C, the embedded epoxy particles (~50 μm) would melt, flow to the crack faces and repair the damage with the help of the excessive amine in the composite. White and his coworkers sealed dicyclopentadiene (DCPD) into microcapsules made from urea-formaldehyde resin. Then, the microencapsulated monomer was added to epoxy-based composites together with powdered Grubbs' catalyst. In the case of cracking, the released DCPD contacts the catalyst, a ring-opening-metathesis polymerization (ROMP) of DCPD would take place and the cracked faces can thus be repaired [30–32]. Furthermore, the methodology was proved to be able to retard and repair fatigue cracks in epoxy composites [33, 34]. To retain the activity of Grubbs' catalyst, it was subsequently encapsulated by wax microspheres so that the catalyst was protected from the influence of amine curing agent in the epoxy resin [35]. Recently, Cho *et al.* presented a self-healing material based on tin-catalyzed polycondensation of phase-separated droplets containing hydroxy end-functionalized polydimethylsiloxane (HOPDMS) and polydiethoxysiloxane (PDES). The catalyst di-*n*-butyltin dilaurate (DBTL) was stored within polyurethane microcapsules embedded in a vinyl ester matrix and released when the capsules were broken by mechanical damage [36].

To simplify the healing techniques based on either resin-filled hollow fibers or microcapsules, Jones *et al.* developed a solid-state healable system, which employs a conventional thermosetting epoxy resin, into which a thermoplastic (i.e. polybisphenol-A-*co*-epichlorohydrin) is dissolved [37, 38]. Upon curing, the thermoplastic material remains dissolved in the thermosetting matrix, in contrast to the conventional thermoplastically toughened matrices. It was hypothesized that upon heating a fractured resin system, the thermoplastic material would mobilize and diffuse through the thermosetting matrix, with some chains bridging closed cracks and thereby facilitating healing.

Considering that most self-healing polymers are inherently electrically insulating, which limits their ultimate responsivities and precludes their use in related analytical applications, Williams *et al.* studied a class of organometallic polymers comprising *N*-heterocyclic carbenes and transition metals as an electrically conductive, self-healing material [39]. These polymers were found to exhibit conductivities of the order of 10^{-3} S cm^{-1} and showed structurally dynamic characteristics in the solid state. Thermal treatment enabled the material to flow and to refill the cracks via a unique depolymerization process.

On the whole, the study on self-healing polymers and polymer composites is still in its infancy. Great efforts have to be made to develop innovative measures and to understand the mechanisms. This chapter reviews the attempts carried out in the authors' laboratory [40], where a two-component healant with the recipe different from those reported previously was proposed, on the basis of the repair approach of microencapsulation [30].

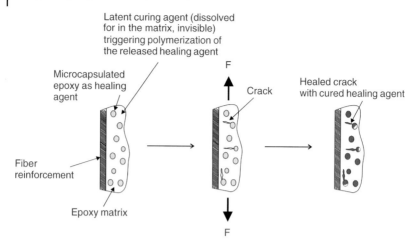

Fig. 2.2 Schematic drawing of the principle of self-healing epoxy-based laminates.

For the purposes of manufacturing self-healing epoxy composites, epoxy oligomer was microencapsulated as polymerizable binder and preembedded in composites' matrix so that miscibility between the healing agent and matrix can be guaranteed. Besides, the complex of $CuBr_2$ and 2-methylimidazole ($CuBr_2(2\text{-MeIm})_4$) was synthesized as the latent hardener of the epoxy healing agent. The complex possesses long-term stability, and can dissociate into $CuBr_2$ and 2-methylimidazole again at around $130-170\,^{\circ}C$ [41–43]. Taking advantage of this characteristic, curing of the released epoxy healing agent catalyzed by 2-methylimidazole (i.e. cracks healing) can be triggered at the dissociation temperature of $CuBr_2(2\text{-MeIm})_4$, which is higher than the curing temperature for making the composites. Another advantage of $CuBr_2(2\text{-MeIm})_4$ lies in its dissolubility in uncured epoxy. As a result, the latent curing agent can be homogeneously predispersed (dissolved) in the composites' matrix on molecular scale. It is believed that this might increase the probability of contact between the epoxy resin from the ruptured microcapsules and the dissociated imidazole. That is, the released epoxy healing agent can be activated wherever it is. The method is more suitable for healing of advanced composites that usually possess higher glass transition temperature. Figure 2.2 shows the concept of the self-healing epoxy-based laminates using the aforesaid healing system.

Although heating is required for starting the healing process, it does not mean manual intervention is necessary. In practical applications, many materials have to work at elevated temperatures. That is, healing of microcracks can be automatically completed as the environmental temperature is high enough to initiate polymerization of the healing agent. On the other hand, with the development of intelligent structural health monitoring [44] or damage self-sensing techniques [45], site-specific heating-induced crack healing can be conducted by accurate positioning of the damages.

2.2
Preparation and Characterization of the Self-healing Agent Consisting of Microencapsulated Epoxy and Latent Curing Agent

2.2.1
Preparation of Epoxy-loaded Microcapsules and the Latent Curing Agent CuBr$_2$(2-MeIm)$_4$

Epoxy healing agent (i.e. bisphenol-A epoxy resin) was microencapsulated by a two-step approach with urea-formaldehyde as the wall material. The typical synthesis procedures are described as follows. First, 20.0 g of urea was mixed with 50.4 ml of formaldehyde (37 wt%) and the pH of the solution was adjusted to 8.0 with 10% NaOH. After reaction for 1 h at 70 °C, transparent water-soluble methylol urea prepolymer was yielded. On the other hand, 40.0 g of epoxy was added to 800 ml of sodium polyacylate (PAANa) solution (1.5 wt%, pH = 8.0) together with resorcinol (4.0 g), NaCl (4.0 g), and polyvinyl alcohol (PVA, 0.8 g). Under the mechanical stirring at a speed of 16 000 rpm, an oil-in-water (O/W) emulsion was formed. Second, the epoxy emulsion was compounded with methylol urea prepolymer to dissolve the latter in the water phase of the former. Afterward, 10% HCl was continuously added to the system by drip feed, and the system was heated (pH 4.0). Eventually, the system was heated to 70 °C while pH reached 2.8–3.0. Having kept at this temperature for 1 h, the system was further adjusted to pH 1.5–2.0. One hour later, the system was neutralized (pH 7) by adding NaOH, cooled down, filtrated, and dried, giving birth to urea-formaldehyde resin-encapsulated epoxy healing agent (Figure 2.3).

Preparation of the latent hardener was relatively simple. CuBr$_2$ (11.2 g) was dissolved in 50 ml of methanol, and then the solution was gradually added to a methanol solution (25 ml) of 2-methylimidazole (16.4 g). Having been stirred for a period of time, the above mixture was diluted by adding 150 ml of acetone, resulting

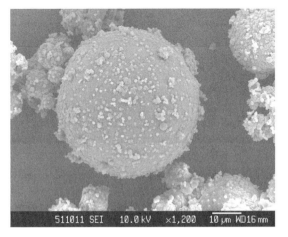

Fig. 2.3 SEM of urea-formaldehyde encapsulated epoxy.

in precipitation. The precipitate (i.e. the complex of $CuBr_2$ and 2-methylimidazole) was filtrated, washed, and dried. The rate of production is around 97.1%.

2.2.2
Characterization of the Microencapsulated Epoxy

The technique of microencapsulation has been developed rapidly since it emerged in the 1950s [46–48]. Although many works were conducted using urea-formaldehyde resin as wall material, the microcapsules containing epoxy has not yet been reported. In our lab, urea and formaldehyde were prepolymerized and then microcapsules were formed via *in situ* condensation as mentioned above. It was found that the pH value of the reaction system exerted critical influence on the products.

When water-soluble methylol urea prepolymer was synthesized in alkaline circumstances, excessive basicity and reaction time led to precipitation of white solids as the methylol urea prepolymer began to be polymerized. It hindered the subsequent *in situ* polymerization in the presence of epoxy emulsion and generated large amount of unwanted cured prepolymers that could be used as the wall material in the ultimate product.

Having been mixed with epoxy emulsion, elimination reaction among methylol urea molecules took place due to the catalysis of alkali or acid. As a result, linear or branched prepolymers with low relative molecular mass were connected through methylene, methylene ether, as well as cyclic bridges between the urea units, evolving into water-insoluble polymer networks with time, depositing onto oil-soluble epoxy droplets, and bringing about microencapsulated epoxy. It is noteworthy that if the condensation polymerization was carried out in an alkaline medium, methylol urea would not react with each other forming methylene linkage, but dimethylene ether linkage. It would lower the functionality of the system, and hence the cross-linking density of the product, which is detrimental to the wall strength of microencapsules. In acidic environment, the product of polycondensation of methylol urea was mainly bonded by methylene linkages, which facilitated chain growth and formation of highly cross-linked structure. Considering that high acidity at the initial stage would result in too high a reaction rate to be controlled, a gradual decrease in pH value was, therefore, followed to prepare the microcapsules containing epoxy.

The Fourier transform infrared (FTIR) spectra of uncured epoxy, microcapsules containing epoxy, and urea-formaldehyde resin that was synthesized under the same conditions as those employed for making the microcapsules are shown in Figure 2.4.

Clearly, strong absorptions appear at 3300–3500 cm^{-1} wavelength of the spectrum of the microcapsules, which represent the stretching modes of –OH and –NH of urea-formaldehyde resin. Besides, the other characteristic peaks of urea-formaldehyde resin, like amine bands (at 1600–1630 cm^{-1} and 1530–1600 cm^{-1}), and those of epoxy including the bands of terminal epoxide group at 914 cm^{-1} and –CH_2– at 2873 and 2929 cm^{-1} are also perceivable from the spectrum

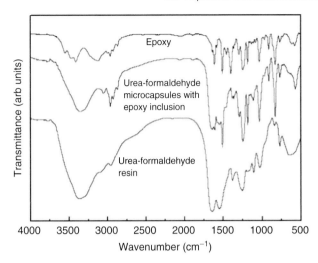

Fig. 2.4 FTIR spectra of bisphenol-A epoxy, urea-formaldehyde encapsulated epoxy, and urea-formaldehyde resin.

of the microcapsules. The results demonstrate that the urea-formaldehyde microcapsules contain epoxy as expected.

The thermal degradation behaviors of the materials give supporting evidence for the above analysis. As illustrated in Figure 2.5, the onset temperature of urea-formaldehyde resin lies in about 242 °C and the maximum rate of pyrolysis appears at 272 °C, while epoxy starts to be degraded at 359 °C and exhibits the maximum decomposition rate at 383 °C.

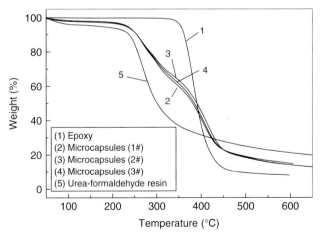

Fig. 2.5 Pyrolytic behaviors of bisphenol-A epoxy, urea-formaldehyde encapsulated epoxy, and urea-formaldehyde resin. Microcapsules (1#–3#) were produced in different batches by the same method. Their average diameters and the contents of the encapsulated epoxy are 36 μm, 64%; 37 μm, 64%; and 31 μm, 61%, respectively.

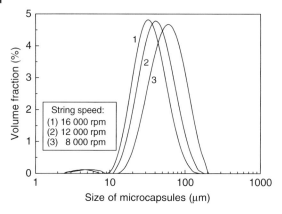

Fig. 2.6 Size distributions of the microcapsules containing epoxy healing agent as a function of stirring speed.

Owing to the thermal degradation of the shell (urea-formaldehyde resin) and the core (epoxy) of the microcapsules, weight loss was observed at maximum rates of pyrolysis (at 266–269 °C and 404–415 °C), as shown in the figure. Compared to the pristine epoxy (curve 1 in Figure 2.5), thermal stability of the epoxy in the microcapsules (curves 2, 3, and 4 in Figure 2.5) evidently increased owing to the protection of the shield of the urea-formaldehyde wall.

Since the epoxy emulsion was produced by high-speed stirring, the ultimate size of the microcapsules should be a function of the stirring speed used for preparing the epoxy emulsion. The plots in Figure 2.6 indicate that with increase in the speed, the microcapsules become smaller and the size distribution narrower.

That is, both size and size distribution is inversely proportional to the stirring speed. This obeys the common knowledge of agitation. In addition, our experimental results showed that the stirring time is another influencing factor. In general, 5 min of stirring is sufficient for the present system to obtain microcapsulated epoxy with stable size.

By weighing the core and shell of the microcapsules (before and after extraction), it was revealed that the microcapsules with different sizes contain different amounts of epoxy (Figure 2.7). The larger microcapsules possess higher amounts of epoxy. This might result from the fact that the thickness of the microcapsules' shells is similar, regardless of their sizes. Therefore, the microcapsules with larger size should have larger portion of the core (i.e. epoxy). On the basis of this relationship, one is able to adjust the concentration of the healing agent.

2.2.3
Curing Kinetics of Epoxy Catalyzed by CuBr$_2$(2-MeIm)$_4$

According to Ref. 42, the possible structure of CuBr$_2$(2-MeIm)$_4$ is shown in Figure 2.8. To verify the synthesized CuBr$_2$(2-MeIm)$_4$, the amount of the nitrogen element in it was measured.

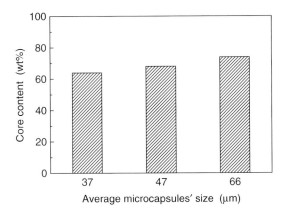

Fig. 2.7 Influence of microcapsules' size on the amount of the epoxy healing agent encapsulated inside urea-formaldehyde shell.

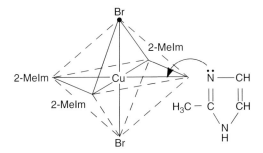

Fig. 2.8 Structure of the latent curing agent $CuBr_2(2\text{-MeIm})_4$.

The value 20.43% is rather close to the theoretical value of 20.31%, suggesting that the target substance has been prepared. Also, the FTIR spectrum of $CuBr_2(2\text{-MeIm})_4$ is compared with 2-MeIm, as shown in Figure 2.9. It is seen that all the characteristic bands of imidazole can be found in the former, including N–H stretching at $3400\,\mathrm{cm^{-1}}$, N–H wagging vibration at $756\,\mathrm{cm^{-1}}$, C–N stretching at $1110\,\mathrm{cm^{-1}}$, C=N stretching at $1600\,\mathrm{cm^{-1}}$, C=C stretching at $1680\,\mathrm{cm^{-1}}$, and =CH rocking vibration at $1440\,\mathrm{cm^{-1}}$.

Therefore, it is evidenced that the imidazole in $CuBr_2(2\text{-MeIm})_4$ retains its original structure as desired. On the other hand, the most significant difference lies in the stretching absorptions of C–H in imidazole ring and methyl at $2500–3200\,\mathrm{cm^{-1}}$, which is not observed in the spectrum of $CuBr_2(2\text{-MeIm})_4$. It manifests that the coordination might have obstructed certain vibration modes.

The mechanism responsible for the reaction between epoxy and a complex of metal salt and imidazole [49] is given in Figure 2.10, which uses $CuCl_2(Im)_4$ (the complex of $CuCl_2$ and imidazole), for example, and is applicable for the system studied in this work.

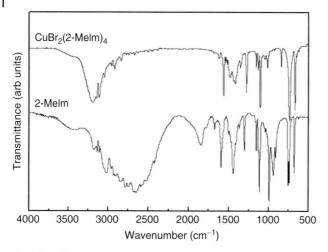

Fig. 2.9 FTIR spectra of 2-Melm and CuBr$_2$(2-Melm)$_4$.

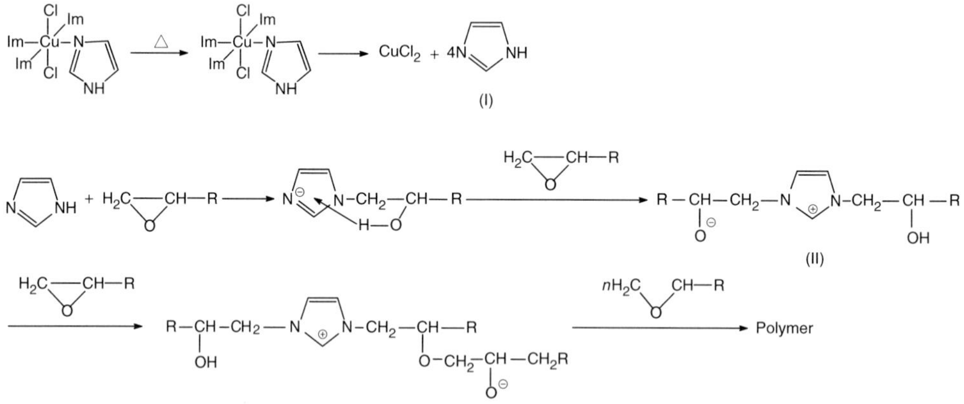

Fig. 2.10 Mechanism of curing reaction between epoxy resin and the complex CuCl$_2$(Im)$_4$.

When heated, the complex dissociates into CuCl$_2$ and imidazole. First, the active hydrogen of the secondary amine in imidazole reacts with an epoxide group yielding affixture. Then, the affixture reacts with another epoxide groupforming an ionic complex. When the anionic portion further reacts with the epoxide group, it can be cured by ring-opening polymerization via a chain reaction. As the anionic polymerization of imidazole is restricted by the affixture, the rate of epoxy curing is lower in comparison to the case where the tertiary amine family is used as the hardener.

Figure 2.11 shows the curing processes of CuBr$_2$(2-Melm)$_4$/epoxy system in terms of conversion versus temperature estimated from the nonisothermal differential scanning calorimetry (DSC) scans (Figure 2.12).

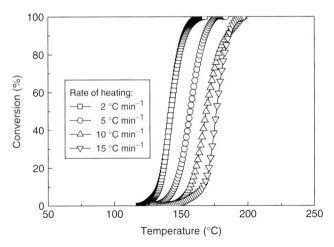

Fig. 2.11 Temperature dependence of conversion of curing reaction of epoxy activated by $CuBr_2(2\text{-MeIm})_4$ (1 wt%) at different heating rates.

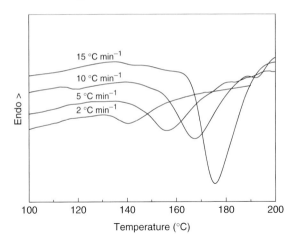

Fig. 2.12 Nonisothermal DSC scans of $CuBr_2(2\text{-MeIm})_4$–epoxy system (1 wt%) at different heating rates.

It is seen that curing of epoxy occurred at about 130 °C. The exothermic peaks appear at 141–176 °C, while the corresponding conversions are below 50%. It means that $CuBr_2(2\text{-MeIm})_4$ is a mild curing agent. By using Kissinger [50] and Crane [51] equations, the characteristic parameters of the curing kinetics, including the activation energy, E_a, and the order of reaction, n, of $CuBr_2(2\text{-MeIm})_4$ (1 wt%)/epoxy, were calculated to be 83.7 kJ mol^{-1} and 0.92, respectively. As n is a nonintegral, the curing of this system should be a complicated process. On the

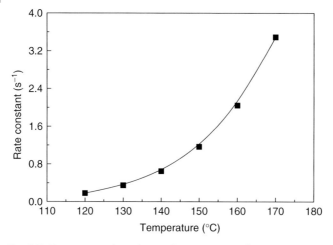

Fig. 2.13 Temperature dependence of rate constant of curing reaction of epoxy activated by $CuBr_2(2\text{-}MeIm)_4$ (1 wt%).

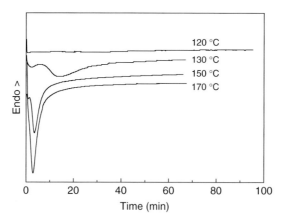

Fig. 2.14 Isothermal DSC scans of $CuBr_2(2\text{-}MeIm)_4$–epoxy system (1 wt%) conducted at different temperatures.

other hand, the temperature dependence of rate constant of the curing reaction was obtained from Arrhenius equation (Figure 2.13).

Clearly, the curing reaction proceeds very slowly at lower temperature (e.g. 120 °C). This is convinced by the isothermal DSC scans of the system (Figure 2.14). At a constant temperature of 120 °C, there is no detectable exothermic peak within 90 min, implying that nearly no curing takes place. In accordance with this finding, the curing temperature for preparing self-healing composites should be lower than the healing temperature, so as to avoid any change in the healing agent and the latent hardener when the composites were cured. In other words, the developed healing system proves to be durable for long-term application under moderate temperature.

2.3
Mechanical Performance and Fracture Toughness of Self-healing Epoxy

2.3.1
Tensile Performance of Self-healing Epoxy

Since the microencapsulated epoxy and the latent curing agent must be filled into the composites' matrix, it is worth understanding their influence on the basic mechanical performance of the matrix. Hence, tensile properties of epoxy containing the self-repair system were measured as a function of the microcapsules' concentration at a fixed content of $CuBr_2(2-MeIm)_4$.

The plots in Figure 2.15a exhibit that tensile strength of the compounds is retained almost unchanged with a rise in the content of the microcapsules. It is different from the results of Brown *et al.* [52], who reported a continuous reduction in the strength of epoxy with embedded microcapsules. Generally, the addition of either rigid particles or rubbery particles into polymers would lead to substantial decrease in tensile strength of the matrices [53, 54]. Nevertheless, counterexamples can be found in inorganic particulates composites where the bonding between the fillers and the matrix is strong enough [55] or the particles' size is in the nanometer range [56]. The strength data in Figure 2.15a suggest that the shell material of the microcapsules, urea-formaldehyde resin, is compatible with epoxy and a strong interfacial interaction was established during curing. Moreover, unlike the soft rubber, the microcapsules are able to carry certain load transferred by the interface. These account for the dependence of tensile strength on the content of microcapsules illustrated in Figure 2.15(a).

The above analysis receives support from the variation trend of Young's modulus (Figure 2.15a).

As the stiffness of the microcapsules ranks between rigid particles and soft rubber, the decrease in modulus of the epoxy specimens is not remarkable, even at a filler concentration as high as 20 wt%. On the other hand, the increase in failure strain (Figure 2.15b) should be attributed to the fact that the microcapsules have induced interfacial viscoelastic deformation and matrix yielding. The decrease in elongation at break for the highly loaded specimens might be due to the uneven distribution of the microcapsules, which led to stress concentration in some parts of the specimen.

2.3.2
Fracture Toughness of Self-healing Epoxy

Similar to the last section, fracture toughness of the epoxy healing agent is also evaluated hereinafter (Figure 2.16). Figure 2.16a indicates that the incorporation of the latent hardener $CuBr_2(2-MeIm)_4$ tends to toughen cured epoxy up to 4 wt%.

It is believed that the latent curing agent dissolved in epoxy plays the role of plasticizer. With increasing content of $CuBr_2(2-MeIm)_4$, agents that cannot be well dissolved have to present themselves in the form of tiny particles (Figure 2.17), and deteriorate the toughening effect.

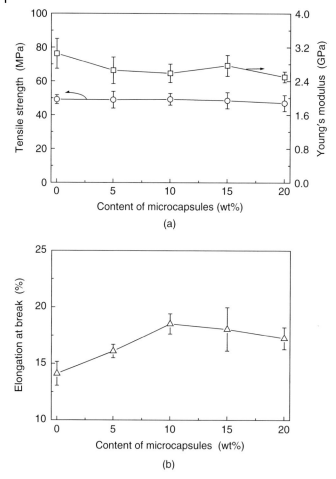

Fig. 2.15 Influence of microcapsules' content on tensile properties ((a) tensile strength and (b) elongation at break) of epoxy. Content of $CuBr_2(2\text{-MeIm})_4$: 2 wt%; average diameter of the microcapsules: 37–42 μm; content of the epoxy healing agent inside the microcapsules: 63–68%.

It explains the reduced toughening efficiency of $CuBr_2(2\text{-MeIm})_4$ from 0.5 to 4 wt%, as compared to the drastic increase in the fracture toughness of the original specimen, K_{IC}^o, when the latent hardener content increases from 0 to 0.5 wt% (Figure 2.16a).

In the case that both microcapsules and the latent curing agent were added, the fracture toughness of the system is slightly lower than that of neat epoxy at certain proportions of the composition. Obviously, the microcapsules can neither hinder the crack propagation nor result in energy consumption. Owing to the positive effect of the latent hardener that counteracted the negative effect of the toughening,

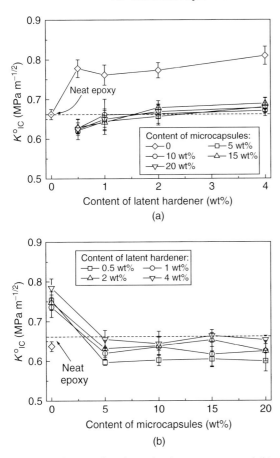

Fig. 2.16 Influence of (a) latent hardener's content and (b) microcapsules' content on fracture toughness of epoxy. Average diameter of the microcapsules: 37–42 µm; content of the epoxy healing agent inside the microcapsules: 63–68%.

fracture toughness of the blends is similar to that of the neat epoxy on the whole (Figure 2.16).

From the above results, it is concluded that the two-component self-healing system developed in the authors' laboratory would not significantly change the mechanical properties of epoxy. Instead, it brings in toughening to some extent in some cases.

2.3.3
Fracture Toughness of Repaired Epoxy

Crack healing of the polymer composites depends on the polymerization of the crack released healing agent (i.e. epoxy in the present work) activated by the latent

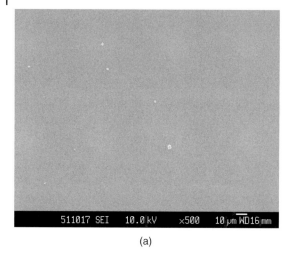

(a)

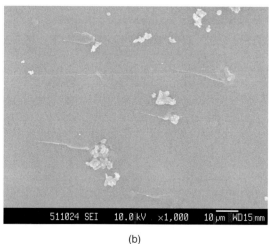

(b)

Fig. 2.17 SEM of cured CuBr$_2$(2-MeIm)$_4$−epoxy system. Content of CuBr$_2$(2-MeIm)$_4$: (a) 0.5 wt% and (b) 2 wt%.

hardener. Since a precise proportion of the ingredients cannot be preestimated for this specific case, a series of materials with different microcapsule–latent hardener ratios have to be prepared for comparing the fracture toughness of the specimens before and after healing (Figures 2.18 and 2.19).

Figure 2.18 shows the dependence of fracture toughness of the healed specimen, K'_{IC}, on the content of the latent hardener. Because the epoxy containing different amounts of the microcapsules and latent hardener possess similar K^o_{IC} (Figure 2.16), the dependence of healing efficiency, η_{epoxy} $\left(\eta_{epoxy} = \frac{K'_{IC}}{K^o_{IC}} \right)$, on the content of the latent hardener is similar to the corresponding curves of K'_{IC} (Figure 2.18).

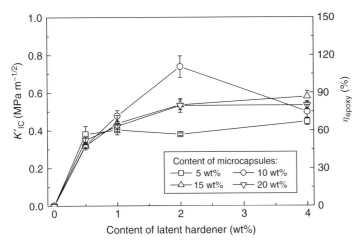

Fig. 2.18 Influence of content of latent hardeneron
self-healing ability of epoxy. Average diameter of the microcapsules: 37–42 μm; content of the epoxy healing agent
inside the microcapsules: 63–68%.

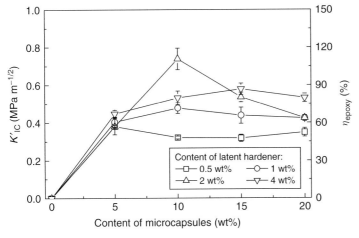

Fig. 2.19 Influence of content of microcapsules on
self-healing ability of epoxy. Average diameter of the microcapsules: 37–42 μm; content of the epoxy healing agent
inside the microcapsules: 63–68%.

When the microcapsules' content is 5 wt%, the values of K'_{IC} are relatively low
within the entire loading range of the latent curing agent. It must be resulting from
the insufficient dosage of the microcapsules leading to insufficient quantity of the
released epoxy for covering the broken surface. Therefore, some cracked portions
will be left unhealed and the apparent fracture toughness is not as high as expected.
In the case of 10 wt% of the microcapsules and 2 wt% of the latent hardener, the
highest K'_{IC} and η_{epoxy} are observed, meaning the optimum proportion has been

reached. Further increase in the content of the latent hardener could not ensure the best cross-linking extent, and therefore the effect of repair has to be lowered accordingly. For the system containing 15 and 20 wt% microcapsules, the areas of the cracked planes that can be healed reduced accordingly. The amount of the latent hardener that might contact the released epoxy becomes insufficient, which also leads to lower repair efficiencies. It is interesting to note that the aforesaid highest healing efficiency is 111%, which implies that the fracture toughness of the healed sample is higher than that of the virgin one. For understanding the cause, fracture toughness of epoxy cured only by $CuBr_2(2\text{-MeIm})_4$ (5 wt%) was measured. The value $0.81\ MPa\ m^{1/2}$ is 1.23 times higher than that of bulky epoxy, which was cured by tetraethylenepentamine. Therefore, the bonding material not only healed the cracks but also provided the damaged sites with higher fracture toughness.

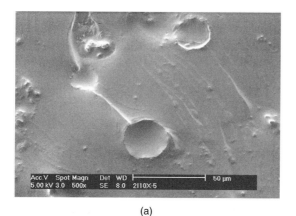

(a)

(b)

Fig. 2.20 SEM of the fractured surfaces of the self-healing epoxy specimens: (a) virgin specimen and (b) healed specimen (the arrow indicates the direction of crack propagation). Contents of microcapsules and $CuBr_2(2\text{-MeIm})_4$: 10 and 2 wt%, respectively.

Figure 2.19 shows the influence of content of microcapsules at certain dosage of the latent hardener. With increasing content of the microcapsules, K'_{IC} increases first, and then decreases.

Besides, the content of the microcapsules corresponding to the maximum K'_{IC} increases with the content of the latent hardener. It demonstrates that the highest healing efficiency can be obtained only at the optimum microcapsule–latent hardener ratio, as seen in Figure 2.18.

The scanning electron microscopy (SEM) of the fractured surface illustrates that after failure of the first single-edge motched bending (SENB) test the microcapsules were damaged and the healing agent had flowed off, leaving ring-like concaves (Figure 2.20a). Having been healed and experienced the second SENB test, the specimen is characterized by more complicated fracture patterns (Figure 2.20b). The thin layers on the fracture surface represent the result of crack propagation through the cured healing agent.

2.4
Evaluation of the Self-healing Woven Glass Fabric/Epoxy Laminates

Woven glass fiber-reinforced polymer composites have been widely used as structural materials in many fields. They are susceptible to damages in the form of delamination generated due to manufacturing defects or low-velocity impact during service. Since the above work has proved that the two-component healant consisting of urea-formaldehyde-walled microcapsules containing epoxy and 2-methylimidazole/metal complex latent hardener is applicable to neat epoxy, the self-healing ability of epoxy-based composite laminates using this healing system should be studied. The systematic knowledge would facilitate optimization of formulation and processing for future practical application of the composites.

In the following investigation, bisphenol-A epoxy also acts as the matrix of the composite laminates. Meanwhile, 0/90 woven glass fabric (C-glass, 13×12 plain weave, 0.2-mm thick, 1-K tows, $200 \, g \, m^{-2}$) is selected to reinforce the composites. The specimens for the investigation are 12-ply laminates with 27 vol% glass fiber, in which the epoxy-loaded microcapsules were preembedded and the latent hardener was predissolved, respectively.

2.4.1
Tensile Performance of the Laminates

Typical tensile stress–strain curves of the composite laminates are shown in Figure 2.21. Evidently, incorporation of the latent hardener into the matrix resin almost does not change the tensile deformation habit (i.e. brittle failure) of the composites (cf. curves 1 and 2 in Figure 2.21).

Before the failure point, the curves are roughly linear at first glance. Nevertheless, the laminates with the preembedded epoxy-loaded microcapsules show

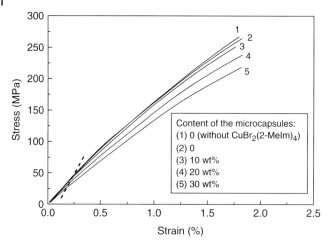

Fig. 2.21 Typical tensile stress–strain curves of the laminates with different contents of the epoxy-loaded microcapsules in the central plies. The intersections between the curves and the dash line roughly indicate the knee points at which the curves deviate from the straight line. Content of $CuBr_2(2\text{-MeIm})_4$: 2 wt%.

slightly different behaviors (curves 3–5 in Figure 2.21). Excessive loading of the microcapsules leads to earlier appearance of the knee point at which the stress–strain curve starts to deviate from linearity. This means that the stable crack propagation in the laminates is closely related to the matrix and fiber/matrix interfacial characteristics in addition to the fact that tensile behavior of the composites is mainly dependent on the fabric reinforcement [57]. The knee points on the stress–strain curves represent the first cracking, and the nonlinear stable propagation between the knee points and the failure points reflects the process of multiple cracking that offers pseudoplasticity for materials [58]. Such pseudoplasticity often reduces the materials' notch sensitivity and improves the reliability. When the cracking starts at the interface between the fibers and the matrix, the microcapsules could behave like stress concentration sites at the interface and lead to the occurrence of knee points at lower load levels with an increased the microcapsule content.

The dependences of Young's modulus and tensile strength on the microcapsules content further reveal the effect of the tiny particles (Figure 2.22). Owing to the lower stiffness of the microcapsules, the rigidity of the matrix has to be decreased, and hence the composites exhibit lower modulus with a rise in the microcapsules fraction. On the other hand, the microcapsules might act as hollow voids in the matrix, which in turn become the sites of stress concentration.

The poor load-bearing capacity of the microcapsules would eventually result in lower strength of the composites. Consequently, tensile strength of the composites with higher microcapsules loading has to be remarkably reduced. Nevertheless, 95% of the strength of the composites excluding the microcapsules can be retained as long as the microcapsules content is lower than 20 wt%. This might

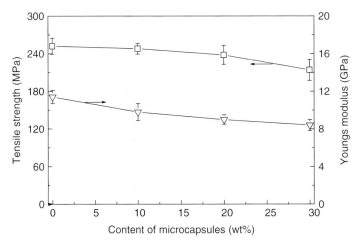

Fig. 2.22 Tensile properties of the laminates as a function of content of the epoxy-loaded microcapsules in the central plies. Content of CuBr$_2$(2-MeIm)$_4$: 2 wt%.

Table 2.1 Porosity of the composites.

Microcapsules content (wt%)	0[a]	10[b]	20[b]	30[b]
Porosity (%)	3.07	3.47	3.73	5.76

a Content of CuBr$_2$(2-MeIm)$_4$: 0.
b Content of CuBr$_2$(2-MeIm)$_4$: 2 wt%.

be related to the change of the matrix resin's viscosity. It was found that the addition of the epoxy-loaded microcapsules at a content lower than 20 wt% only slightly influenced the viscosity of the laminating resin. However, further addition of the microcapsules drastically increased the resin's viscosity. It must severely deteriorate the impregnation and the interfacial coupling as is evidenced by the greatly increased porosity of the laminates at 30 wt% of the microcapsules (Table 2.1). Therefore, the tensile performance of the laminates became worse.

2.4.2
Interlaminar Fracture Toughness Properties of the Laminates

To evaluate the self-healing capability of the composites, double cantilever beam (DCB) tests were conducted. Healing was conducted by keeping the tested specimen in an oven preset at 130 °C for 1 h. Then, the healed specimen was taken out of the oven and rested at room temperature for 24 h. Finally, the healed sample was tested again in terms of DCB configuration. Figure 2.23 shows the typical load-crack opening displacement curves of the specimens. For the virgin laminates, the

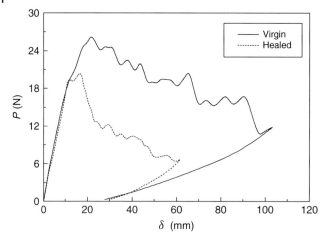

Fig. 2.23 Typical load (P)–crack opening displacement (δ) curves of the self-healing laminates before and after healing (note: unloading curves are not shown). Contents of the epoxy-loaded microcapsules and $CuBr_2(2\text{-MeIm})_4$ in the central plies: matrix epoxy/microcapsules/$CuBr_2(2\text{-MeIm})_4$ $= 100/20/2$.

behavior is approximately linear up to the onset of crack growth near the peak load of about 27 N.

Upon further displacement, the crack propagates along the midplane in a stable mode and the load falls slowly. Through seven load–unload circulations, that is, 70 mm crack extension, the crack opening displacement attains 105 mm. In contrast, the healed specimen closely follows the original loading curve until the crack begins to advance again near the load of 20 N. Then, the load drops rapidly along with the crack growth, which is quite unstable compared to the virgin crack propagation. When the crack length reaches 70 mm, the crack opening displacement is only 65 mm. That is, delamination resisting ability of the healed midplane is reduced.

The analysis receives support from SEM of the fractured surfaces of the composites (Figures 2.24 and 2.25). Figure 2.24a shows that after failure due to the first DCB test fiber, breakage and debonding are obvious, besides the rupture of the epoxy-loaded microcapsules. These damage modes must have consumed substantial energy, resulting in higher interlaminar performance. The magnified view of the resin-rich region (Figure 2.24b) further indicates that the healing agent had flowed out of the broken microcapsules as expected, leaving ring-like concaves.

After the second DCB test, the repaired specimen exhibits complicated fracture patterns. Figure 2.25a illustrates that the main areas that have been repaired should be the resin-rich region. The cured healing agent has formed thin layers on the fractured surface (Figure 2.25b). It is worth noting that the quantity of the released healing agent might be insufficient to build up a compact coverage

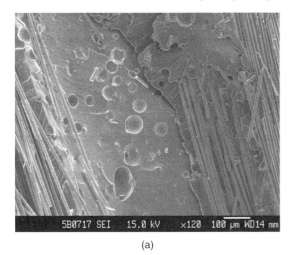

(a)

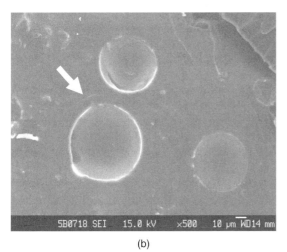

(b)

Fig. 2.24 SEM of the fractured surfaces of the self-healing laminates (obtained after the first DCB test). The arrow indicates the wave front of the released epoxy. Contents of the epoxy-loaded microcapsules and $CuBr_2(2\text{-MeIm})_4$: 30 and 2 wt%, respectively.

on the fiber-rich region. In addition, the broken fibers cannot be reconnected by the present self-healing agent. Consequently, the contribution of fiber bridging to hinder crack development has to be lowered.

In Figure 2.26, the typical delamination resistance curves (R-curves) of the composite laminates are presented. Mode I critical strain energy release rate, G_{IC}, was calculated according to the modified beam theory (MBT) method. For the virgin laminates with and without the self-healing agent, the tendencies of their R-curves and G_{IC} values are similar. G_{IC} increases with crack propagation until it reaches a

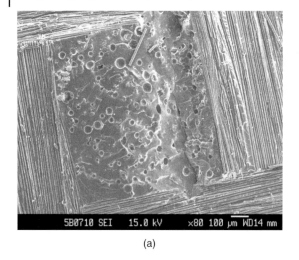

(a)

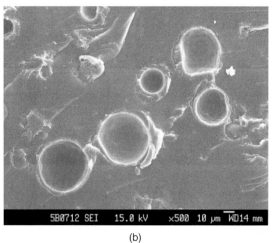

(b)

Fig. 2.25 SEM of the fractured surfaces of the self-healing laminates (obtained after the second DCB test). Contents of the epoxy-loaded microcapsules and $CuBr_2(2\text{-MeIm})_4$: 30 and 2 wt%, respectively.

steady state after the extension of crack about 10 mm. The plateau values are about twice the initial values.

The results imply that the preembedded microcapsules and predissolved catalyst nearly have no effect on interlaminar fracture toughness of the composites because fiber bridging is the main influencing factor during the first DCB tests. For the healed specimen, its G_{IC} declines after reaching the maximum until the values become roughly the same as the initial values. Clearly, lack of in-plane fiber bridging in front of the crack should account for the phenomenon. The above estimation is evidenced by the morphological observations (Figure 2.25). That is, the open areas

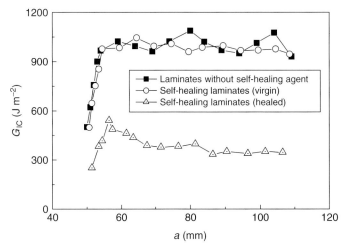

Fig. 2.26 Interlaminar fracture toughness of the laminates, G_{IC}, as a function of crack length, a. Contents of the epoxy-loaded microcapsules and $CuBr_2(2\text{-Melm})_4$ in the central plies of the self-healing laminates: matrix epoxy/microcapsules/$CuBr_2(2\text{-Melm})_4 = 100/20/2$.

of the weave were rich in the epoxy-loaded microcapsules probably due the fact that the microcapsules were filtered to the specific sites during compression molding, so that the healing took place mostly in the open weave areas. The repaired portions failed to provide sufficiently strong closure force as a result.

Table 2.2 systematically compares the effects of concentration of the microcapsules and catalyst and thickness of the laminates on G_{IC} and healing efficiency. It is seen that critical strain energy release rate in the virgin specimen, G_{IC}^{Virgin}, decreases with increasing content of the epoxy-loaded microcapsules, while the influence of the catalyst content is negligible.

This might be due to the fact that the presence of the epoxy-loaded microcapsules raised the matrix viscosity, and hence deteriorated impregnation. The worse interfacial adhesion would facilitate crack propagation. However, with respect to the average and maximum critical strain energy release rate in the healed specimens, $G_{IC}^{Healed,avg}$ and $G_{IC}^{Healed,max}$, the catalyst content has more significant influence than the microcapsules content. For the laminates with 20 wt% microencapsulated epoxy, for example, both $G_{IC}^{Healed,avg}$ and $G_{IC}^{Healed,max}$ increase with increasing catalyst concentration from 0.5 to 2 wt%. Accordingly, the maximum healing efficiency of the composites is raised by nearly two times. It is interesting to see from Table 2.2 that a further increase in the dosage of the catalyst at constant microcapsules content could not continuously increase the healing efficiency. Similarly, dependence of the self-healing efficiency on the microcapsules content also varies in this way. Therefore it is obvious that the microcapsules-to-catalyst ratio is a key factor that determines the self-healing efficiency of the laminates. If the concentrations of the microcapsule and catalyst were mismatched, satisfactory repair result could not be

Table 2.2 Dependence of healing efficiency on the influencing factors.

Variables			$G_{IC}^{Virgin,avg}$ (J m^{-2})	G_{IC}^{Healed} (J m^{-2})		η^a (%)	
Microcapsules (wt%)	Catalyst (wt%)	Thickness (mm)		Avg.	Max.	Avg.	Max.
0	0	4.3	1059 (±137)	0	0	0	0
20	0.5	4.5	905 (±165)	102 (±56)	138 (±119)	34	39
	1		931 (±110)	185 (±81)	327 (±85)	45	59
	2		950 (±112)	206 (±38)	478 (±154)	47	71
	3		927 (±142)	198 (±55)	431 (±162)	46	68
10			967 (±187)	135 (±65)	236 (±213)	37	49
30	2	4.5	838 (±151)	205 (±83)	517 (±123)	49	79
35			799 (±138)	211 (±114)	446 (±105)	51	75
10			942 (±100)	116 (±93)	238 (±199)	35	50
30	3	4.5	856 (±170)	197 (±78)	489 (±97)	48	76
35			787 (±154)	203 (±109)	442 (±141)	51	78
20	2	4.0	989 (±162)	84 (±57)	154 (±145)	29	39
		4.8	935 (±136)	154 (±124)	225 (±198)	41	49

a η: Crack healing efficiency $\left(\eta = \dfrac{K_{IC}^{Healed}}{K_{IC}^{Virgin}} = \sqrt{\dfrac{G_{IC}^{Healed}}{G_{IC}^{Virgin}}} \right)$.

acquired. The data in Table 2.2 reveal that the highest healing efficiency is perceived at 30 wt% microcapsules and 2 wt% catalyst. It implies that higher loading of the microcapsules is generally required for delivering sufficient polymerizable healing agent to the cracked portions so as to obtain higher healing efficiency. However, it is worth noting that a rise in the microcapsules concentration would lead to significant reduction in tensile properties of the laminates (Figures 2.21 and 2.22). Therefore, a balance between strength and toughness restoration should be considered when manufacturing the self-healing composites for practical application.

To confirm that the healing at 130 °C happens only because of the microcapsules plus the latent hardener, and not because of the unreacted epoxy monomer contained within the host laminates, FTIR spectra of the fractured surfaces of the self-healing laminates before healing and after heat treatment that simulated the healing condition were collected, in comparison with that of the laminates without the self-healing agent. As shown in Figure 2.27, the band of the terminal epoxide group appears at 914 cm^{-1} of the spectrum of the fracture surface of the self-healing composites (curve 2), while it is it is not observed for the laminates without the self-healing agent (curve 1). It proves that there is no unreacted epoxy in the host laminates and the broken microcapsules are able to provide uncured epoxy on the fracture surface. Having been heated at 130 °C for 1 h, this epoxide peak disappears (curve 3), indicating that the epoxy from the broken microcapsules have been consolidated by the latent hardener.

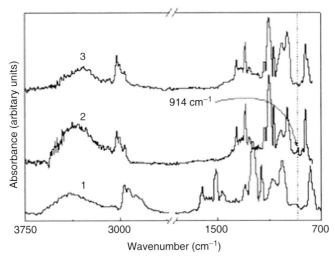

Fig. 2.27 FTIR spectra of fractured surface of (1) the laminates without the self-healing agent, (2) the laminates with the self-healing agent (contents of the epoxy-loaded microcapsules and $CuBr_2(2\text{-Melm})_4$ are 20 and 2 wt%, respectively), and (3) the specimen (2) heated at 130 °C for 1 h simulating the healing procedure.

The stronger absorptions of C–O–C stretching mode (at 1100–1140 cm^{-1}) and the weaker absorptions of –OH stretching mode (at 3300–3500 cm^{-1}) on curve 2 than those on curve 3 support the above analysis. Therefore, it can be concluded that the preembedded epoxy-loaded microcapsules and the latent hardener are the only resources responsible for the crack healing.

In fact, thickness of the laminates is also indispensable for the efficiency of self-healing. As exhibited in Table 2.2, the thinner specimens (4 mm in thickness) offer lower healing efficiency. This should be related to the processing of the laminates. That is, the microcapsules might have burst during the compression molding and the effective amount of the microcapsules in the composites had to be reduced. Even in the case of thicker specimens (4.8 mm in thickness), the healing efficiency is not high. Removal of the air bubbles deep inside the composites by degasification was to some extent difficult. The healed areas might be reduced, and hence the healing efficiency too. Only when the laminates have optimum thickness (4.5 mm, for instance), higher healing efficiency can be yielded. Further study in this aspect will be conducted in the subsequent work.

2.4.3
Self-healing of Impact Damage in the Laminates

Woven fabric reinforced thermosetting composites are widely used as structural components in many fields. Low-velocity impact damages are often generated during their service [59, 60]. These damages (like matrix microcracking and interfacial debonding) generally are on the microscopic level and invisible. When the microcracks propagate, mechanical performance of the materials rapidly declines, particularly the compression strength [61]. For example, damage resulting from impact can cause a loss of 50% of compressive strength. Therefore, impact damage is one of the primary factors for the design of structural composites [62].

The discussion in the last section shows the ability of the aforesaid homemade self-healing agent in recovering interlaminar fracture toughness of epoxy laminates. Therefore, it is worth exploring its capability of healing impact damage in the laminates. Compression after impact (CAI), a measure of damage tolerance of composite laminates, is employed as the main characterization means hereinafter. To check the self-healing capability of the composites, the specimens were first impacted and then healed at 140 °C for 0.5 h under (i) no added pressure or (ii) under 60 or (iii) 240 kPa, respectively. Figure 2.28 shows the flowchart of the tests.

Low-velocity impact of fiber-reinforced plastics has been the subject of many experimental and analytical investigations, for example [60–73]. Impact energy and laminate layup are the crucial factors determining the damage modes of composites, which include matrix cracking, delamination, and fiber fracture.

Figure 2.29 shows the morphologies of the impact damage zones taken from the rear surface of the composite laminates subjected to different impact energies. Slight matrix cracking at low impact energy level of 1.5 J is the main damage mode (Figure 2.29a and b). With a rise in impact energy, the visibility of damage size extends and the damage mode becomes more complicated. It is consistent

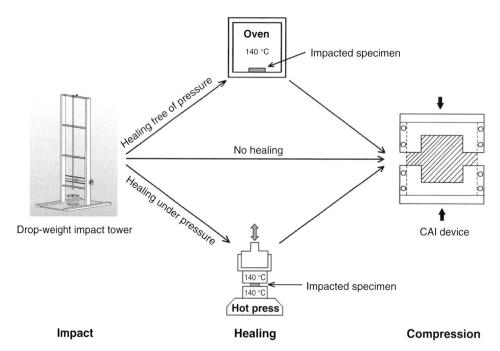

Fig. 2.28 Flowchart of the damage healing procedures and CAI tests.

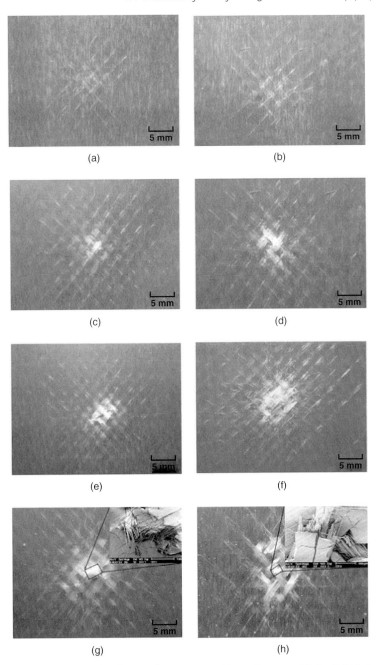

Fig. 2.29 Photographs showing the impact damage zones on the woven glass fabric/epoxy composite laminates with and without the self-healing agent. Content and size of the microencapsulated epoxy in the self-healing composites: 10 wt% and 40 μm. Content of CuBr$_2$(2-MeIm)$_4$ in the self-healing composites: 2 wt%. Impact energy: (a, b) 1.5 J; (c, d) 2.0 J; (e, f) 2.5 J; (g, h) 3.5 J. The insets in (g) and (h) show the broken fibers.

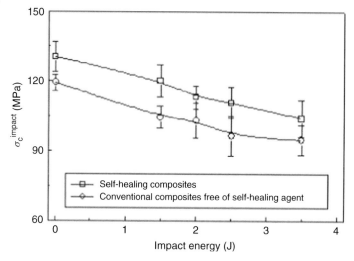

Fig. 2.30 Residual compressive strength of the composite laminates as a function of impact energy. Content and size of the microencapsulated epoxy in the self-healing composites: 10 wt% and 40 µm. Content of CuBr$_2$(2-MeIm)$_4$ in the self-healing composites: 2 wt%.

with the results of Hiral *et al.* [68]. In Figure 2.29c and d, substantial matrix cracking appears in the immediate vicinity of the impact contact area, and initial delamination is also perceivable. At 2.5 J impact, there occurs obvious deformation on the unimpacted part owing to the growth of delamination and slight fiber rupture (Figure 2.29e and f). At a higher impact energy level of 3.5 J, the dominant damage mechanisms are extensive delamination and fiber fracture (Figure 2.29g and h and the insets).

Compared to the composites containing the self-healing agent, the composites free of self-healing agent were severely damaged under the same impact condition. It suggests that the microencapsulated epoxy and the latent hardener might have helped in absorbing some impact energy.

Effect of impact energy on residual compressive strength of the composite laminates with and without self-healing agent is shown in Figure 2.30. Incorporation of the self-healing agent obviously enhanced the compressive strength of the composites. This is evidenced by the fact that the curve for the composites with healing agent has higher value than that of the conventional composites at all impact energy levels. It might result from the presence of the self-healing agent that is strongly adhered to the matrix epoxy, which brings about additional constraint and increases the resistance to compressive instability. The residual compressive strength of the impacted composites decreases with increasing impact energy, which matches the increase in the damage area [61, 68].

To highlight the influence of impact energy on the residual compressive strength of the composites in an objective way, the strength data of the impacted specimens are normalized with respect to that of the virgin ones. As shown in Figure 2.31,

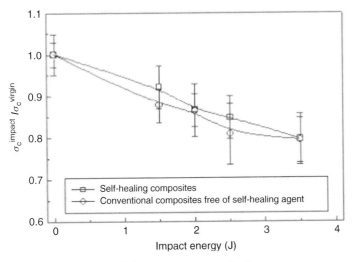

Fig. 2.31 Normalized residual compressive strength of the composite laminates as a function of impact energy. Content and size of the microencapsulated epoxy in the self-healing composites: 10 wt% and 40 µm. Content of $CuBr_2(2\text{-}Melm)_4$ in the self-healing composites: 2 wt%.

introduction of self-healing agent into the composites slightly retarded the declining rate of the residual compressive strength with impact energy.

The result coincides with the variation in both damage area and damage mode of the composites showed in Figure 2.29

Compressive strength of the healed impacted laminates is plotted as a function of impact energy in Figure 2.32. Besides, the healing efficiency of the specimens is also given in Figure 2.33 to quantify the effect of crack repair. At 1.5 J impact energy, the compressive strength of the impacted specimen after healing procedures, σ_c^{healed}, is very close to the compressive strength of the virgin specimen, σ_c^{virgin}, with healing efficiencies from 89 to 94% depending upon the healing conditions.

The above results mean that the cracks in the matrix of the laminates were almost completely healed. Consequently, the damage zone is no longer visible in the T-scan ultrasonic images (Figure 2.34a–c), which indicates zero damage area in the repaired laminates.

In fact, the healing of the cracked matrix can be further evidenced by the simulative indentation experiment. As shown in Figure 2.35, having been healed in an oven, the split in the resin-rich region of the edge-cracked laminates is filled with consolidated healing agent that binds the cracked planes together. The porous structure of the seam might result from the pressure-free healing procedure.

When impact energy is increased, both σ_c^{healed} and η_{CAI} are reduced. The main reason might be due to the change of the impact damage mode. At low impact energy level, the predominant damage characterized by matrix cracking is on the microscopic scale, and so the released healing agent can easily reach the cracked

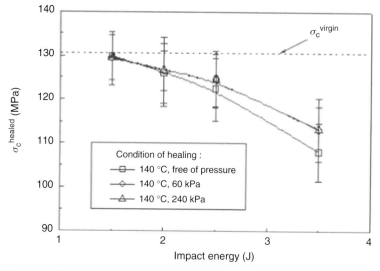

Fig. 2.32 Influence of impact strength on residual compressive strength of the repaired impacted composite laminates. Content and size of the microencapsulated epoxy in the self-healing composites: 10 wt% and 40 μm. Content of $CuBr_2(2\text{-MeIm})_4$ in the self-healing composites: 2 wt%.

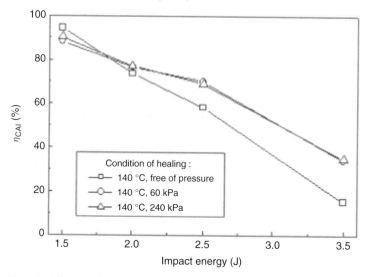

Fig. 2.33 Influence of impact energy on the healing efficiency, η_{CAI}, of CAI specimens under different healing conditions. Content and size of the microencapsulated epoxy in the self-healing composites: 10 wt% and 40 μm. Content of $CuBr_2(2\text{-MeIm})_4$ in the self-healing composites: 2 wt%.

$$\eta_{CAI} = \frac{\sigma_c^{healed} - \sigma_c^{impact}}{\sigma_c^{virgin} - \sigma_c^{impact}} \times 100\%,$$ where σ_c^{healed} denotes the compressive strength of the impacted specimen after healing procedures, σ_c^{impact} is the compressive strength of the impacted specimen, and σ_c^{virgin} is the compressive strength of the virgin specimen.

| Impacted | Repaired free of pressure | Repaired (pressure: 240 kPa) |

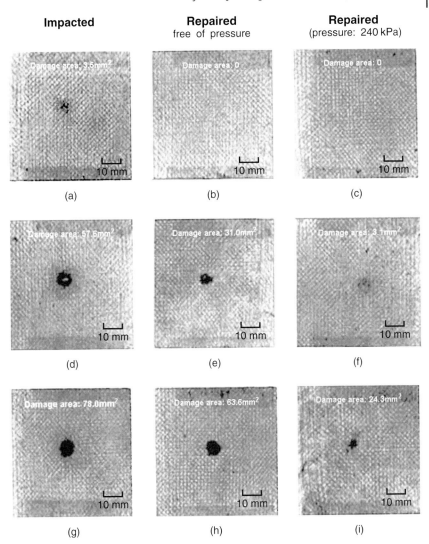

Fig. 2.34 T-scan ultrasonic images of the impacted self-healing composite laminates before and after being healed. Content and size of the microencapsulated epoxy: 10 wt% and 40 µm. Content of $CuBr_2(2\text{-MeIm})_4$: 2 wt%. Impact energy: (a–c) 1.5 J; (d–f) 2.5 J; and (g–i) 3.5 J.

places for rehabilitation. In the case of high impact energy level, the damage mode is dominated by delamination and fiber breaking resulting in less effective repair of the cracks (Figure 2.34d–i). First, the healing agent is unable to reconnect the broken fibers [74]. Second, delivery of enough healing agent to the delaminated sites deep inside the fiber-rich regions might be difficult. As a result, the healing efficiency has to be low.

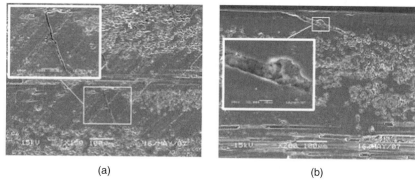

(a) (b)

Fig. 2.35 SEM side views of (a) indented composite laminates and (b) indented composite laminates healed in
an oven at 140 °C for 0.5 h. Content and size of the microencapsulated epoxy: 10 wt% and 40 μm. Content of
CuBr$_2$(2-Melm)$_4$: 2 wt%.

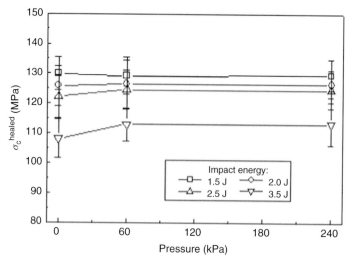

Fig. 2.36 Influence of pressure on σ_c^{healed} of the composite
laminates. Content and size of the microencapsulated epoxy:
10 wt% and 40 μm. Content of CuBr$_2$(2-Melm)$_4$: 2 wt%.

The above analysis received support from the dependence of σ_c^{healed} on pressure
(Figure 2.36) that was applied to the impacted specimens during healing treatment
(Figure 2.28). The aid of external pressure can narrow the cracked gaps during
the course of repairing, so that the amount of the self-healing agent required for
filling up the splits would be less as compared to the healing, free of pressure. As
the impact damage is directly related to the impact energy (Figure 2.34), effect of
the pressure on healing efficiency is more prominent in the laminates subjected to

high impact energy. Therefore, when the pressure is raised from 0 to 60 kPa, the increment of σ_c^{healed} of the laminates experienced 3.5 J impact is greater than that of the laminates experienced 2.5 J impact (Figure 2.36).

On the other hand, Figure 2.36 indicates that σ_c^{healed} of all the composite laminates remains unchanged with a rise in pressure from 60 to 240 kPa. It reveals that the external pressure is irrelative to the mechanism of healing, and only relative to the closing the two crack surfaces. It is interesting to note that σ_c^{healed} of the specimens subjected to 1.5 J impact has nothing to do with the pressure. Evidently, absence of severe damages such as delamination and fiber fracture should explain the phenomenon. It can thus be concluded that applying pressure on the specimens during healing is necessary for the composites subjected to high-energy impact to improve the efficiency of repairing.

In fact, failure modes of the CAI specimens are greatly affected by the healing. As illustrated in Figure 2.37a, failure of the impacted composite laminates is primarily caused by the crack propagating from the existing damage on the back face [68].

The fracture plane is approximately perpendicular to the direction of compression (Figure 2.37b). When the impacted specimen is healed and then subjected to compression, shear buckling takes place similar to conventional laminates [75]. The fracture plane makes an angle of about 45° with the loading direction and the broken parts become noncoplanar (Figure 2.37c and d), implying that the compression failure starts with the kink zone formation [76] that propagates into the fracture. It manifests that the impact damage must have been recovered to a great extent before the compression test. In addition, Figure 2.37c indicates that the compression failure site is not located in the original impact damage zone (i.e. the central region) of the healed specimen, which is different from the case shown in Figure 2.37a. The mechanism is unclear. As the average fracture toughness of the bulk epoxy cured by $CuBr_2(2\text{-MeIm})_4$ (i.e. the cured self-healing agent) is 0.81 MPa·m$^{1/2}$, which is 1.76 times higher than that of the bulk epoxy cured by 2-ethyl-4-methylimidazole (i.e. the cured matrix of the laminates), it might be inferred that the higher fracture toughness of the healed portion than the matrix leads to higher resistance to compressive failure of the former. Consequently, the ultimate compression failure has to occur in a place other than the healed portion. Further study is needed to prove this estimation.

Effect of content of the self-healing agent is showed in Table 2.3. When the composite laminates do not contain any self-healing agent, healing of the impacted specimen has no influence on its compressive strength, so the values of σ_c^{impact} and σ_c^{healed} are identical and $\eta_{CAI} \approx 0$. It proves that the host matrix resin of the composites has no self-healing capability.

The data in Table 2.3 further reveals that the healing efficiency is improved as the content of the microencapsulated epoxy is increased from 10 to 20 wt% at a fixed loading of $CuBr_2(2\text{-MeIm})_4$. This should be a result of the increased amount of the released epoxy, which ensures that much more crack volumes can be filled up by the binder.

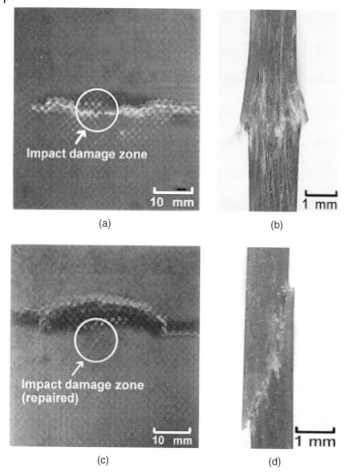

(a) (b)

(c) (d)

Fig. 2.37 (a and b) Photographs of composite laminates that were impacted at 2.0 J and then compressed to failure. (c, d) Composite laminates that were impacted at 2.0 J, repaired in a hot press under 60 kPa at 140 °C for 0.5 h, and then compressed to failure. (a and c) front views and (b and d) side views. Content and size of the microencapsulated epoxy: 10 wt% and 40 μm. Content of $CuBr_2(2\text{-MeIm})_4$: 2 wt%.

Besides the content of the self-healing agent, the size of the microcapsules containing epoxy also affects healing efficiency. At a given content of the self-healing agent, the healing efficiency slightly increased on increasing the microcapsules' size from 40 to 140 μm (Figure 2.38). Similar findings were reported elsewhere [52, 77]. Here, because the larger capsules contain more epoxy (Table 2.4), their breakage will deliver more healing agent to the cracked surfaces as compared to the small capsules.

From Table 2.3, it is known that 10 wt% microencapsulated epoxy (40 μm) is insufficient for healing cracks. Therefore, an increase in the microcapsules' size would have the same effect as an increase in the amount of available healing agent.

Table 2.3 Compressive strengths and healing efficiencies of the composite laminates[a].

Content of microencapsulated epoxy[b] (wt%)	σ_c^{virgin} (MPa)	Impacted at 1.5 J			Impacted at 2.5 J		
		σ_c^{impact} (MPa)	$\sigma_c^{healed\,c}$ (MPa)	η_{CAI} (%)	σ_c^{impact} (MPa)	$\sigma_c^{healed\,c}$ (MPa)	η_{CAI} (%)
0[d]	119.3 ± 3.5	104.6 ± 4.8	105.2 ± 5.5	∼0	96.4 ± 8.7	95.5 ± 7.2	∼0
10	130.5 ± 6.5	120.2 ± 6.8	129.9 ± 5.5	94	110.8 ± 6.7	122.3 ± 7.2	58
20	127.3 ± 5.2	120.9 ± 5.9	127.6 ± 3.3	100	106.3 ± 4.2	121.3 ± 3.2	71

a Content of $CuBr_2(2\text{-MeIm})_4$: 2 wt%.
b Average diameter of the microcapsules: 40 μm.
c Conditions of crack repairing: pressure free, 140 °C, and 0.5 h.
d Composite laminates free of the microencapsulated epoxy and the latent hardener.

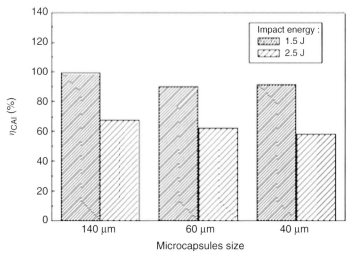

Fig. 2.38 Influence of size of the microcapsules containing epoxy on the self-healing efficiency of CAI specimens. Contents of the microencapsulated epoxy and $CuBr_2(2\text{-MeIm})_4$: 10 and 2 wt%, respectively. Conditions of crack repairing: pressure free, 140 °C, and 0.5 h.

Table 2.4 Specifications of the epoxy-loaded microcapsules.

Average diameter (μm)	140	65	40
Core content (wt%)	81.2	73.9	63.9

2.5
Conclusions

Self-healing of cracks in epoxy-based composites in terms of preembedded microencapsulated epoxy and predissolved latent curing agent proves to be a prospective approach. The healing efficiency is a function of many factors, including the size of the microcapsule, amount of the healing agent, and even the degree of damage. In case the hardener of the two-component healing agent system has sufficiently high activity at or below room temperature, activation of the healing process can be conducted within a much wider temperature range. This should be the way forward for the healant.

Although achievements in the field of self-healing polymers and polymer composites are far from satisfactory, the new opportunities that were found during research and development have demonstrated that it is a challenging job to either invent new polymers with inherent crack repair capability or integrate existing materials with novel healing systems. Interdisciplinary studies based on close collaboration among scientists are prerequisites for overcoming the difficulties. Comparatively, the techniques using healing agent might be advantageous for industrialization. The works and outcomes in this aspect have broadened the application possibility of polymeric materials. Also, the extended service life of components made from these intelligent materials would contribute to reduce waste disposal. Toohey and coworkers [78], for instance, constructed self-healing materials with microvascular networks capable of repairing repeated damage events. It is undoubtedly important for building up a sustainable society. From a long-term perspective, synthesis of brand-new polymers accompanied by self-healing function through molecular design would be a reasonable solution. Recent exploration has shown the prospects of this trend, but the automatic trigger mechanism remains open.

Acknowledgments

The authors are grateful to the support of the Natural Science Foundation of China (Grants: 50573093, U0634001).

References

1 Riefsnider, K.L., Schulte, K. and Duke, J.C. (1983) *ASTM STP*, **813**, 136–59.

2 Trask, R.S., Williams, H.R. and Bond, I.P. (2007) *Bioinspiration and Biomimetics*, **2**, 1–9.

3 Jud, K. and Kausch, H.H. (1979) *Polymer Bulletin*, **1**, 697–707.

4 Kausch, H.H. and Jud, K. (1982) *Plastics and Rubber Processing and Applications*, **2**, 265–68.

5 Wool, R.P. and O'Connor, K.M. (1981) *Journal of Applied Physics*, **54**, 5953–63.

6 Kim, Y.H. and Wool, R.P. (1983) *Macromolecules*, **16**, 1115–20.

7 Wool, R.P. (1982) *ACS Polymer Preprints*, **23** (2), 62–63.

8 Jud, K., Kausch, H.H. and Williams, J.G. (1981) *Journal of Materials Science*, **16**, 204–10.

9 McGarel, O.J. and Wool, R.P. (**1987**) *Journal of Polymer Science Part B-Polymer Physics*, **25**, 2541–60.

10 Wool, R.P. and Rockhill, A.T. (**1980**) *ACS Polymer Preprints*, **21**, 223–24.

11 Lin, C.B., Lee, S. and Liu, K.S. (**1990**) *Polymer Engineering and Science*, **30**, 1399–406.

12 Wang, E.P., Lee, S. and Harmon, J. (**1994**) *Journal of Polymer Science Part B-Polymer Physics*, **32**, 1217–27.

13 Takeda, K., Unno, H. and Zhang, M. (**2004**) *Journal of Applied Polymer Science*, **93**, 920–26.

14 Takeda, K., Tanahashi, M. and Unno, H. (**2003**) *Science and Technology of Advanced Materials*, **4**, 435–44.

15 Chen, X.X., Dam, M.A., Ono, K., Mal, A., Shen, H., Nutt, S.R., Sheran, K. and Wudl, F. (**2002**) *Science*, **295**, 1698–702.

16 Chen, X.X., Wudl, F., Mal, A.K., Shen, H.B. and Nutt, S.R. (**2003**) *Macromolecules*, **36**, 1802–807.

17 Dry, C. (**1992**) *International Journal of Modern Physics B*, **6**, 2763–71.

18 Dry, C. (**1996**) *Composite Structures*, **35**, 263–69.

19 Zhao, X.P., Zhou, B.L., Luo, C.R., Wang, J.H. and Liu, J.W. (**1996**) *Chinese Journal of Materials Research (in Chinese)*, **10** (1), 101–4.

20 Motuku, M., Janowski, C.M. and Vaidya, U.K. (**1999**) *Smart Materials and Structures*, **8**, 623–38.

21 Bleay, S.M., Loader, C.B., Hawyes, V.J., Humberstone, L. and Curtis, P.T. (**2001**) *Composites Part A*, **32**, 1767–76.

22 Trask, R.S., Williams, G.J. and Bond, I.P. (**2007**) *Journal of the Royal Society Interface*, **4**, 363–71.

23 Pang, J.W.C. and Bond, I.P. (**2005**) *Composites Part A*, **36**, 183–88.

24 Pang, J.W.C. and Bond, I.P. (**2005**) *Composites Science and Technology*, **65**, 1791–99.

25 Trask, R.S. and Bond, I.P. (**2006**) *Smart Materials and Structures*, **15**, 704–10.

26 Williams, G., Trask, R.S. and Bond, I.P. (**2007**) *Composites Part A*, **38**, 1525–32.

27 Jung, D., Hegeman, A., Sottos, N.R., Geubelle, P.H. and White, S.R. (**1997**) in *Proceedings of American Society for Mechanical Engineers (ASME), Symposium on Composites and Functionally Graded Materials*, Vol. MD-80 (K., Jacob, N. Katsube and W., Jones), ASME, Dallas, TX, pp. 265–75.

28 White, S.R., Sottos, N.R., Geubelle, P.H., Moore, J.S., Seriram, S.R., Kessler, M.R. and Brown, E.N. (**2005**) U.S. Patent 6,858,659 B2, Feb 22, 2005.

29 Zako, M. and Takano, N. (**1999**) *Intelligent Journal of Material Systems and Structures*, **10**, 836–41.

30 White, S.R., Sottos, N.R., Geubelle, P.H., Moore, J.S., Kessler, M.R., Seriram, S.R., Brown, E.N. and Viswanathan, S. (**2001**) *Nature*, **409**, 794–97.

31 Brown, E.N., Sottos, N.R. and White, S.R. (**2002**) *Experimental Mechanics*, **42** (4), 372–79.

32 Kessler, M.R., Sottos, N.R. and White, S.R. (**2003**) *Composites Part A*, **34**, 743–53.

33 Brown, E.N., White, S.R. and Sottos, N.R. (**2005**) *Composites Science and Technology*, **65**, 2466–73.

34 Brown, E.N., White, S.R. and Sottos, N.R. (**2005**) *Composites Science and Technology*, **65**, 2474–80.

35 Rule, J.D., Brown, E.N., Sottos, N.R., White, S.R. and Moore, J.S. (**2005**) *Advanced Materials*, **17**, 205–80.

36 Cho, S.H., Andersson, H.M., White, S.R., Sottos, N.R. and Braun, P.V. (**2006**) *Advanced Materials*, **18**, 997–1000.

37 Hayes, S.A., Jones, F.R., Marshiya, K. and Zhang, W. (**2007**) *Composites Part A*, **38**, 1116–20.

38 Jones, F.R., Hayes, S.A. and Zhang, W. (**2007**) *Journal of the Royal Society Interface*, **4**, 381–87.

39 Williams, A., Boydston, J. and Bielawski, W. (**2007**) *Journal of the Royal Society Interface*, **4**, 359–62.

40 Yin, T., Rong, M.Z., Zhang, M.Q. and Yang, G.C. (**2007**) *Composites Science and Technology*, **67**, 201–12.

41 Dowbenko, R., Anderson, C.C. and Chang, W.H. (**1971**) *Industrial and Engineering Chemistry Product Research and Development*, **10**, 344–51.

42 Ibonai, M. and Kuramochi, T. (**1975**) *Purasuchikkusu (in Japanese)*, **26** (7), 69–73.

43 Bi, C.H., Gan, C.L. and Zhao, S.Q. (**1997**) *Thermosetting Resin (in Chinese)*, **12** (1), 12–15.

44 Su, Z. and Ye, L. (**2004**) *Smart Materials and Structures*, **13**, 957–69.

45 Fernando, G.F., Degamber, B., Wang, L.W., Doyle, C., Kister, G. and Ralph, B. (**2004**) *Advanced Composites Letters*, **13**, 123–29.

46 Gardner, G.L. (**1966**) *Chemical Engineering Progress*, **62** (4), 87–91.

47 Fanger, G.O. (**1974**) *Chemtech*, **7**, 397–405.

48 Song, J., Chen, L. and Li, X.J. (**2001**) *Microencapsulation and its Application (in Chinese)*, Chemical Engineering Press, Beijing.

49 Eilbeck, W.J., Holmes, F., Taylor, C.E. and Underhill, A.E. (**1968**) *Journal of the Chemical Society (A)*, **1**, 128–32.

50 Kissinger, H.E. (**1957**) *Analytical Chemistry*, **29**, 1702–6.

51 Crane, L.W., Dynes, P.J. and Kaelble, D.H. (**1973**) *Journal of Polymer Science Polymer Letters Edition*, **11**, 533–40.

52 Brown, E.N., White, S.R. and Sottos, N.R. (**2004**) *Journal of Materials Science*, **39**, 1703–10.

53 Ahmed, A. and Jones, F.R. (**1990**) *Journal of Materials Science*, **25**, 4933–42.

54 Kinloch, A.J. and Young, R.J. (**1983**) *Fracture Behaviour of Polymers*, Applied Science Publishers, London.

55 Jancar, J., Dianselmo, A. and Dibenedetto, A.T. (**1992**) *Polymer Engineering and Science*, **32**, 1394–99.

56 Rong, M.Z., Zhang, M.Q., Zheng, Y.X., Zeng, H.M., Walter, R. and Friedrich, K. (**2000**) *Journal of Materials Science Letters*, **19**, 1159–61.

57 Hull, D. (**1981**) *An Introduction to Composite Materials*, Cambridge University Press, New York.

58 Leung, C.K.Y. and Li, V.C. (**1990**) *Composites*, **21**, 305–17.

59 Horton, R.E. and McCarty, J.E. (**1987**) Damage tolerance of composites, in *Engineered Materials Handbook, Vol.1, Composites* (ed. T.J. Reinhart), ASM International, Metals Park, Ohio, pp. 259–67.

60 Soutis, C. and Curtis, P.T. (**1996**) *Composites Science and Technology*, **56**, 677–84.

61 Pritchard, J.C. and Hogg, P.J. (**1990**) *Composites Part A-Applied Science*, **21**, 503–11.

62 Baker, A.A., Jones, R. and Callinan, R.J. (**1985**) *Composite Structures*, **4**, 15–44.

63 Saez, S.S., Barbero, E., Zaera, R. and Navarro, C. (**2005**) *Composites Science and Technology*, **65**, 1911–19.

64 Bogdanovich, A.E. and Friedrich, K. (**1994**) *Composite Structures*, **27**, 439–56.

65 Davies, G.A.O., Hitchings, D. and Zhou, G. (**1996**) *Composites Part A-Applied Science*, **27**, 1147–56.

66 Shen, W.Q. (**1997**) *International Journal of Impact Engineering*, **19**, 207–29.

67 Naik, N.K. and Sekher, Y.C. (**1998**) *Journal of Reinforced Plastics and Composites*, **17**, 1232–63.

68 Hiral, Y., Hamada, H. and Kim, J.K. (**1998**) *Composites Science and Technology*, **58**, 91–104.

69 Naik, N.K., Sekher, Y.C. and Meduri, S. (**2000**) *Composites Science and Technology*, **60**, 731–44.

70 Kim, J.K. and Sham, M.L. (**2000**) *Composites Science and Technology*, **60**, 745–61.

71 Cartié, D. and Irving, P. (**2002**) *Composites Part A-Applied Science*, **33**, 483–93.

72 Baucom, J.N. and Zikry, M.A. (**2005**) *Composites Part A-Applied Science*, **36**, 658–64.

73 Baucom, J.N., Zikry, M.A. and Rajendran, A.M. (**2006**) *Composites Science and Technology*, **66**, 1229–38.

74 Chen, X.B. (**2004**) *Polymer-Matrix Composites Handbook (in Chinese)*, Chemical Industry Press, Beijing.

75 Chou, T.W. (**1993**) *Structure and Properties of Composites*, Wiley-VCH Verlag GmbH, Weihheim.

76 Carlsson, L.A. and Pipes, R.B. (**1997**) *Experimental Characterization of Advanced Composite Materials*, Technomic Publishing, Lancaster.

77 Rule, J.D., Sottos, N.R. and White, S.R. (**2007**) *Polymer*, **48**, 3520–29.

78 Toohey, K.S., Sottos, N.R., Lewis, J.A., Moore, J.S. and White, S.R. (**2007**) *Nature Materials*, **6**, 581–85.

3

Self-Healing Ionomers

Stephen J. Kalista, Jr.

3.1
Introduction

> *Of all the self-healing systems, self-healing ionomers are able to autonomously recover from the most devastating damage in a very short period of time, and are by far the least expensive to manufacture.*
> —Bergman and Wudl [1]

Imagine yourself standing 1 m from a window holding a loaded firearm. Upon pulling the trigger, one might expect the glass to shatter, perhaps showering the ground with a multitude of glass shards. However, when you fire, the bullet goes through and you strain to see the results. In fact, upon approaching the window, you find that the hole once there has healed itself leaving only a small scar indicating the point of bullet penetration. Such a feat would seem like a magic trick or the realm of science fiction; however, it is exactly what would happen if the window were fabricated from the ionomer materials presented in this chapter. This unique self-healing response has been recently observed in a certain class of thermoplastic poly(ethylene-co-methacrylic acid) (EMAA) ionomers as an automatic and instantaneous self-repair following ballistic puncture [2–4]. The repair of damage is complete and fully sealed to leakage even under significant pressure [3, 5–7]. An example of the observed ballistic healing phenomenon is given in Figure 3.1.

To date, self-healing research has focused predominantly on creating systems that heal through chemical means. Those designed by White *et al.* [9], Pang and Bond [10, 11], and others have quite elegantly addressed the self-repair of damage, utilizing embedded microspheres or tubes that contain a healing agent. Upon injury, these components release the healing agent into the surrounding matrix allowing successful repair of damage, including fatigue cracks, delaminations, or similar modes [9–11]. However, they would not be expected to heal following such a macroscopic and energetic damage event as ballistic puncture. Ionomer self-repair is indeed very different and unique. In fact, the self-healing ability is an inherent material response rather than by design, suggesting a very different

Self-healing Materials: Fundamentals, Design Strategies, and Applications. Edited by Swapan Kumar Ghosh
Copyright © 2009 WILEY-VCH Verlag GmbH & Co. KGaA, Weinheim
ISBN: 978-3-527-31829-2

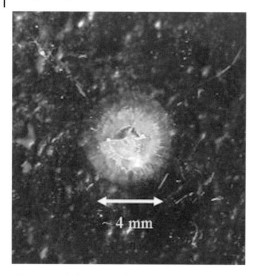

Fig. 3.1 Healed ionomer film following ballistic puncture [8] (reprinted with permission by Springer Science and Business Media).

healing mechanism. It occurs automatically and instantaneously without manual intervention—indeed it is an autonomic self-healing process. Given this, much research has focused on uncovering the mechanism of self-healing in these materials and in determining when this self-healing behavior might occur. Such understanding is necessary to design applications that utilize this unique response. This chapter provides a review of the work on self-healing ionomers, the mechanism behind this ability, and the methods used to uncover this unique response, including ballistic and other techniques. It probes both the range of situations in which healing will occur and the development of novel self-healing ionomer composites using this technology. Finally, it reveals why self-healing is not simply an ionic phenomenon but rather the behavior of a class of copolymers and related ionomers which exhibit this unique autonomic healing response.

3.2
Ionomer Background

Although ionomers have been around for over 40 years [12], research on their unique fundamental properties is still being conducted. Since inception, their definition has grown and changed with the research body. Initially, the term *ionomer* described only the original ionic group containing olefin-based polymers of Rees and Vaughan [12]. As more ion-containing polymers were introduced, an update was needed and the definition morphed to fit. In their work, Tant and Wilkes [13] defined ionomers as a class of ion-containing copolymers in which the maximum ion group content was ~15 mol% (a definition still commonly used today). However, given the lack of distinction between such systems and polyelectrolytes, further revision was necessary. With a clearer picture of the unique morphological character of these materials and its effect on properties, Eisenberg and Rinaudo [14] ultimately defined ionomers on a functional rather than compositional basis. Under their definition, ionomers were defined as polymers whose bulk properties are governed by ionic interactions within discrete regions of the polymer structure. The characteristic at the core of this definition is the ionic aggregate, which is discussed in some detail in the subsequent text. Such gradual change in definition certainly reflects the continuous study of the unique chemical structure, morphology, and physical properties presented by these materials. Indeed, the interaction between these components is still debated in the literature, though all indicates a wealth of potential in the ionomer field [15]. Given the discovery and early findings of self-healing in at least a group of these materials [2, 3, 6], much focus is sure to follow. To understand the self-healing behavior, the reader should first understand the unique morphological characteristics of ionomers.

3.2.1
Morphology

Several texts address the field of ionomer science in detail and the specific character of these fascinating materials [15, 16]. Ionomers are commonly produced via a unique copolymer neutralization process. Beginning with precursor copolymers containing both ionic (anionic) and nonionic repeat groups, the anionic acid component is neutralized. This neutralization process provides ionic character by the formation of an ionic pair with an associated metallic cation (the counterion). This produces an ionomer (or ionic copolymer) with pendant ionic groups attached along the polymer chain. Indeed, many variations are possible whereby these ionic groups can be distributed randomly along the backbone, at chain ends, or in prescribed fashions leading to random or even block copolymer architectures depending on the copolymer synthesis and resulting structure. Additionally, given changes in the anionic groups or counterion chosen, other structural differences would result. Of course, as this neutralization process yields the ionic content, a change in the fraction of acid groups present in the initial copolymer and/or the number of these groups that are neutralized will cause the amount of ionic content to vary.

Given the unique attractions provided by the presence of these ionic groups, ionomers are noted to possess a very unique and interesting microstructural character. The ionic pairs present in these materials have been shown to group into discrete regions known as *multiplets*. Eisenberg [17] defined a multiplet as an aggregate consisting of several ion pairs, the number of which is limited by steric effects of the adjacent polymer chain segments and the size of the ion pair. Further, according to the Eisenberg–Hird–Moore (EHM) model [18], these ion pairs anchor their attached polymer chains to the multiplet. This provides a sort of physical cross-link in the polymer structure considerably reducing mobility of the attached polymer chains in that vicinity. Owing to this reduced mobility compared to that of the bulk polymer, the surrounding area is known as the *restricted mobility region*. A representation of this concept is given in Figure 3.2. In the figure shown,

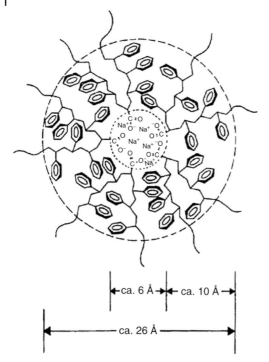

Fig. 3.2 EHM model of the ionic multiplet and surrounding restricted mobility region in poly(styrene-*co*-sodium methacrylate) [18] (reprinted with permission by the American Chemical Society).

the multiplet is the region within the inner dashed circle that contains the ionic pairs. Surrounding the multiplet is the restricted mobility region consisting of the attached polymer chains and extending to the outer dashed circle. Beyond its limits, the polymer chains return to their typical bulk mobility. From this model, a clear understanding of the multiplet is obtained. (Note: the ionomer structure used in Figure 3.2 provides an example of an ionic multiplet; however, it does not represent a known self-healing ionomer material.)

The next larger step in the ionomer hierarchy described by Eisenberg *et al.* [18] is the ionic cluster. An increase in ionic content creates numerous multiplets within the ionomer structure. As ionic content increases, multiplet density increases, therefore the restricted mobility regions of the neighboring multiplets begin to overlap forming a more continuous restricted mobility region throughout the polymer structure. This new continuous region is defined as the ionic cluster and acts as a second phase within the ionomer, even expressing its own T_g. This concept is described in Figure 3.3. In the figure, three regions "a", "b", and "c" are shown, containing differing amounts of ionic content. The multiplets are indicated as small circles with the surrounding restricted mobility region of each represented as a gray area. As the level of ionic content is increased from low to high, the restricted mobility regions are shown to overlap, producing these clusters.

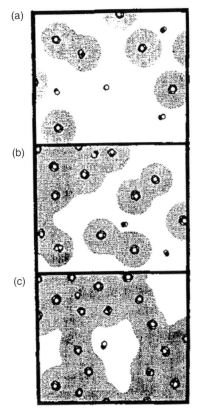

Fig. 3.3 Schematic image representing ionic cluster formation with increase in ionic content (from a to c) [18] (reprinted in part with permission by the American Chemical Society).

The significant role of ionic content in determining the unique microstructure is well reflected in ionomer behavior. It is firmly established in the literature that the amount of ionic content has a drastic effect on mechanical properties [19–22]. Bellinger *et al.* [19] noted sodium sulfonated polystyrene ionomers (Na-SPS) to show an increase of ~60% in tensile strength and ~100% in toughness with an increase from 0 to ~7.5 mol% ionic content. For EMAA materials, Statz [20] and Rees [21] showed a two- to fivefold increase in modulus for ionic acid neutralization levels up to 40%. Further, Rees also noted a near 50% increase in tensile strength when neutralization of EMAA was raised from 0 to 80%. Though these are only a few examples, they serve to demonstrate the significant effect of ionic aggregation and clustering.

The unique ionic character is also noted to impact thermal characteristics. Tadano *et al.* [23] showed that the ionic clusters exhibit a first-order transition during heating. This transition was located below the onset of the T_m of the crystalline regions of the polymer chain and was identified as an order–disorder

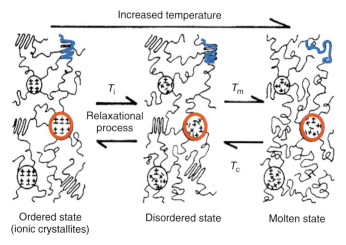

Increased temperature

T_i

Relaxational process

T_m

T_c

Ordered state
(ionic crystallites)

Disordered state

Molten state

Fig. 3.4 Model representing the effect of the heating/cooling cycle on ionomer order–disorder and crystallization [23] (reprinted with permission by the American Chemical Society).

transition during which the ionic groups disorder when heated above T_i. Upon subsequent cooling, it was shown that the ionic groups then reordered over a long time relaxation process with the T_i peak observed to advance to higher temperatures and increase in definition rather slowly. For the EMAA materials studied here, the relaxation process was noted to take ~38 days for the peak to revert back to its original size and location. Figure 3.4 summarizes these thermal events schematically. From the left-hand side, the ionic groups start in the ordered state as circled in red. Upon increased temperature, they pass through T_i and the ionic regions disorder. With further heating, above T_m, the polymer crystallites are noted to melt (blue). Subsequently, upon cooling through the T_c, polymer crystallites re-form. When reaching the cooled, room temperature (RT) state, the ionic regions remain disordered only to reorder through the long-term relaxation process described above while effectively annealing at RT.

As noted above, the ionic character of these materials plays a significant role in controlling their unique structural nature. It provides them with novel thermal characteristics and produces significantly altered mechanical properties compared to the copolymers from which they are formed. Given the unique structure of ionomers and the significant effect of ionic aggregation on thermomechanical properties, it was expected that ionic content and its unique inter/intramolecular attractions were responsible for their novel self-healing ability [2].

3.2.2
Ionomers Studied for Self-healing

Several polymers from the thermoplastic EMAA family have been assessed for the self-healing behavior [2–8]. With respect to the ionomer synthesis discussed

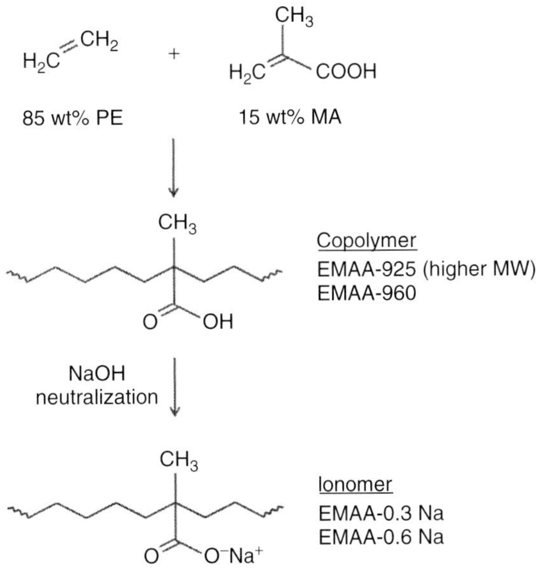

Fig. 3.5 Synthesis and structure of EMAA ionomers and nonionic copolymers assessed for self-healing.

above, the original random EMAA copolymer contains 15 wt% (5.4 mol%) anionic methacrylic acid groups (Figure 3.5). Two types of these nonionic copolymers have been studied. These include DuPont Nucrel 925 and Nucrel 960 [24], which will be referred to in this text as EMAA-925 and EMAA-960. Two ionomer materials formed from neutralization of the precursor EMAA copolymer have also been examined. These include DuPont Surlyn 8940 with 30% of its acid groups neutralized and Surlyn 8920 with 60% neutralized [25]. They are referred to as EMAA-0.3Na and EMAA-0.6Na in the text. All four types have melt transition temperatures between 91 and 93 °C [5, 6]. In general, the EMAA materials are known to have a high clarity, excellent toughness, and favorable cut, abrasion, and chemical resistance. Because of these properties, they have found use in coatings, packaging, and in a range of sporting equipment [2]. By examining numerous ionic and nonionic materials from the same family, it was believed that much could be learned about the effect of ionic content and structure on the self-healing response [2, 3, 6].

3.3
Self-healing of Ionomers

Ionomer self-healing has been studied by only a handful of researchers and to date the published literature is still sparse. It includes three peer-reviewed articles [5, 6, 26], three theses [2, 3, 27], several conference proceedings and other documents [4, 7, 8, 28–31], and one chapter [32]. The early history of the discovery of ballistic healing in ionomers is not well documented in the open literature. According to Huber and Hinkley [28], a 1996 patent specifies DuPont Surlyn 8940 (EMAA-0.3Na) as a material to be used in the fabrication of "self-sealing" shooting range targets [33]. The cited advantage of such targets was their potential for use following multiple impacts with a much longer life versus those of paper, wood, or fiberboard. The patent, filed on October 22, 1993, is the first known and documented discussion of self-healing in ionomers. Fall [2] reported the existence of a commercial product, which is presumed to be based on this patent. React-A-Seal, produced by Reactive Target Systems, Inc., is an ionomer product based on DuPont Surlyn 8940 and is marketed for the aforementioned purpose.

The ballistic self-healing behavior could potentially be utilized in many applications beyond shooting range targets. Several studies suggest use as a self-healing barrier material in space vessels and structures subject to matter impacts during spaceflight [2, 5, 6, 28]. Similarly, Coughlin *et al.* [31] examined its potential for use in navy aircraft fuel tanks and fluid-containing parts, allowing improved craft survivability following combat damage. One might imagine potential use in other areas as diverse as a self-sealing barrier for chemical containment or in medical applications whereby a thin self-healing membrane may prove beneficial. The discovery of a repair response beyond that of ballistic puncture healing would yield an increased potential for use in these and other applications. In order to effectively utilize these self-healing materials or to engineer and synthesize new polymers that might do the same or heal in other cases, the mechanism and capabilities of self-healing in EMAA materials must be well understood.

3.3.1
Healing versus Self-healing

A scholarly search for the term *healing* will yield any number of articles unrelated to the self-healing behavior. Beyond those discussing biological systems, there are many dealing with thermoplastic materials, including ionomers. However, the term *healing* is separate and distinct from *self-healing*, and it involves no damage event.

When two polymer surfaces are placed in intimate contact above their T_g, the surfaces will begin to bond together. Though there is no damage occurring, this autohesion or interfacial welding phenomenon has commonly been referred to in the literature as *healing* [34, 35]. It occurs as polymer chains interdiffuse across the polymer–polymer interface via polymer chain reptation. This model of chain motion has been described quite elegantly, particularly in the work of de Gennes [36], though others have also addressed it. Such a process of interfacial healing has been studied extensively for polymers placed in contact above T_g (for amorphous) or above T_m (for semicrystalline polymers). Given the mobility of the polymer chains at these temperatures, such motion would not be unexpected. However, healing has also been observed at sub-T_g conditions for other polymers, including polystyrene and polyethylene terephthalate [37]. As would be anticipated for any of these cases, bond strength has been observed to increase with elevated healing temperature and greater contact pressure. Although the above concept of interfacial healing is interesting, the process occurs in a wide array of polymers (e.g. low-density polyethylene (LDPE), polystyrene (PS), and terephthalate (PET)) that exhibit no tendency toward a self-healing response. However, one can transfer these concepts to the case of puncture healing in EMAA ionomers. As the bullet passes through an EMAA film, it must create a split (interface) in the material. Therefore, the theories presented in the interfacial healing literature may still be quite important in understanding the self-healing behavior. In many cases in this chapter, the term *healing* will be used interchangeably with *self-healing*; however, when the interfacial healing phenomenon is implied, the distinction will be noted.

Self-healing can be distinguished as a very different phenomenon, which is a material response to a damage event. As mentioned previously, self-healing will occur in different systems (ionomers or otherwise) in response to various damage modes. The predominant self-healing response for EMAA ionomers is that following ballistic puncture damage. During an opening lecture at the First International Conference on Self-Healing Materials, White defined '*self-healing*' as "the ability to recover functionality after damage using the inherent resources of a materials system" [38]. This behavior could be specifically characterized as "autonomic" if it were achieved in an "independent and automatic fashion" [38]. Though not engineered to do so, the self-healing behavior of the EMAA system fits both of these definitions. However, much is unknown about this behavior and when it will occur. A more complete understanding of various damage modes eliciting healing is presented in the following section.

3.3.2
Damage Modes

So far, ionomer self-healing has been characterized as a response following puncture damage. However, other modes have yielded a healing behavior. Kalista [3, 5] provided a survey of several of these cases. While simple cutting with scissors or razor blade did not elicit healing, the more energetic damage mode of sawing did. In work by Siochi and Bernd [39], films of the React-a-Seal material (ie based on EMAA-0.3Na) were cut manually with a hacksaw. The frictional cutting process generated heat in the material, though it was not sufficient to provide significant repair. However, cutting with a heat gun present (80–90 °C) provided the thermal energy necessary to produce a healing response (Figure 3.6). Such a scenario has many parallels to the interfacial healing behavior described above and suggests melt or near-melt heating as requisite to self-repair. Puncture methods were also examined. Though nail puncture has similar geometry to ballistic impact and would be quite energetic and likely to generate significant heat, it did not exhibit self-healing upon nail removal [3, 5]. It was concluded that the continued presence of the nail even for a few seconds allowed the polymer to relax, producing plastic drawing. This suggested the requirement of an elastic component to the healing mechanism for ballistic self-repair.

Ballistic self-healing of ionomers has been studied quite extensively [2–6]. EMAA films (typically ~6 mm and thinner) have shown the ability to repair when impacted by bullets and similar projectiles for a range of velocities. Wound closure occurs in a practically instantaneous fashion such that high speed imaging at 4000 frames

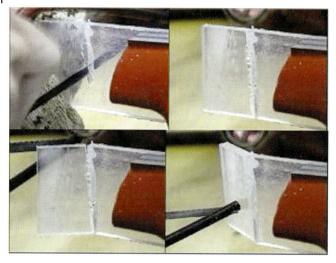

Fig. 3.6 Time-lapse photos showing healing upon saw-ing damage. Images progress chronologically from the top left in a clockwise fashion (reproduced from Kalista and Ward [5]).

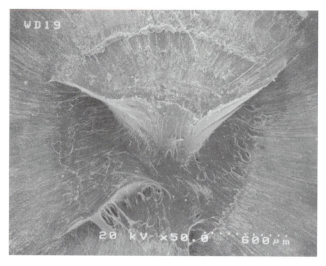

Fig. 3.7 Scanning electron micrograph of healed EMAA-0.3Na film following puncture with a 4.5-mm-diameter projectile [42].

per second has proved inconclusive in capturing the event [6]. Following shooting, only a smaller healed scar remains on the sample (Figures 3.1 and 3.7) with the damage site sealed to pressurized air [3, 5, 6]. Further studies have concluded EMAA self-healing to be repeatable following multiple ballistic punctures [40], while Varley and van der Zwaag [41] also observed successful repeatability in a nonballistic test which is discussed in the subsequent text.

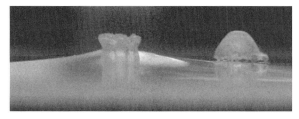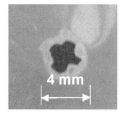

Fig. 3.8 Ballistic puncture damage in LDPE films (reproduced from Kalista et al. [6] www.informaworld.com).

Varley has described three discrete regions visible in scanning electron microscopy (SEM), which include an outer region of radially oriented striations, a smooth region, and finally a crater at the center [32]. The outer two regions are visible in Figure 3.7, though the crater is not as prominent as it was in the work of Varley. This might suggest that the higher caliber projectiles used by Varley produced more devastating damage leaving a larger crater. The self-healing response of EMAA is also quite different from that of LDPE, which experienced a brittle and melt elongational fracture upon puncture leaving the damaged piece, as shown in Figure 3.8.

Self-healing is typically studied using pointed projectile shapes (with consistent healing), though other geometries have yielded varying results. An exhaustive study of shape has not been conducted, but the results indicate interesting information about the healing process. Flat-headed projectiles were observed to remove polymer from the film leaving a circular punched hole [2], while similar lower momentum projectiles became lodged in the film with the polymer tightly squeezing the pellet neck between head and tail [40]. Spherical BBs have also been found to become lodged in the film upon impact. In this case, the BB had passed the plane of the film, though it was encapsulated in a thin layer that conformed to its spherical shape. A hole remained on the entrance side, it though had contracted to a size smaller than that of the projectile diameter [40]. Such studies lend additional support to the possibility of a significant elastic component of the self-healing behavior.

Thermal response upon damage was also noted. Using a thermal IR camera, Fall [2] noted temperature to rise into the melt during damage (~98 °C) for a wide range of projectile speeds upon puncture with a 9-mm bullet. As distance from the center of the puncture site increased, the temperature quickly dropped to that of the surroundings. Differential scanning calorimetry (DSC) performed at the damage site by Kalista and Ward [5] also confirmed a rise in the temperature above the melt upon damage. The size of the melt region was also shown to be quite small, on the order of the projectile or smaller [5, 40].

3.3.3
Ballistic Self-healing Mechanism

Kalista *et al.* proposed a two-stage process as the mechanism for ionomer ballistic self-healing [3, 6]. Such a scheme is described in Figure 3.9. The self-healing behavior occurs upon projectile puncture whereby energy must be transferred to

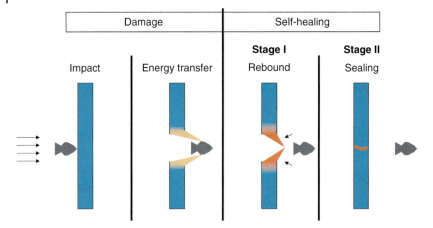

Fig. 3.9 Two-stage model for ballistic self-healing in EMAA films.

the material during impact, both elastically (stored) and inelastically (dissipated as heat through friction). Upon impact, the local puncture site is heated to the melt state through friction and extension of the ionomer film. The presence of ionic groups provides an enhanced mechanism for heating and elastic energy storage in the melt. The film then fails allowing the projectile to pass through. Using the cooler, more rigid perimeter area around the puncture site as a frame, the polymer responds with sufficient melt elasticity to rebound and close the puncture (stage I). The molten surfaces then bond together and interdiffuse in order to seal the hole on a very short time scale (stage II). As film puncture creates an interface, this second stage can be related to the aforementioned interfacial healing phenomenon, and its presence is confirmed by the significant pressures held by the film after the damage event. The two main requirements expected necessary to produce the self-healing behavior are that the puncture event must produce a local melt state in the polymer material and that the molten material must have sufficient melt elasticity to snap back and close the hole.

Similar to stage II, Kalista and Ward [5] have suggested a potential third stage of long-term strengthening. Stage III is expected to involve continued interdiffusion, crystallization, and long-term relaxations of the polymer chains and in particular of the ionic structures. Although it is not necessary to produce the self-healing response, it would strengthen the healed site with continued aging. Future study should address such a process.

3.3.4
Is Self-healing an Ionic Phenomenon? (Part I)

While a model for self-healing is established above, the reason *why* these materials heal is still unanswered. Fall [2] suggested healing would occur if sufficient energy was transferred to the polymer during impact, heating it above the T_i and disordering the aggregates. Upon projectile exit and local cooling, the ionic aggregates would reorder causing the ionomer to heal. This provided an ionic theory for the self-healing response. Given the unique intermolecular attractions and the significant effect of the ionic morphology on properties, such a theory seemed reasonable. To test the theory, Fall examined EMAA-0.3Na, EMAA-0.6Na, and React-a-Seal ionomers and the nonionic EMAA-925. Upon ballistic puncture (9-mm bullets), it was observed that all samples indicated a healing response with an apparent closing of the hole. Additionally, a water droplet placed on the damage site did not pass through, seemingly confirming Fall's healing assessment.

In subsequent laboratory testing (air rifle, 4.5-mm pointed pellets), Kalista examined the ionic phenomenon in greater detail [3, 6]. Here, the two ionic materials (EMAA-0.6Na and EMAA-0.3Na) and the two nonionic copolymers (EMAA-925 and EMAA-960) were examined. After observing that even a water droplet would not pass through films that had small but obvious holes owing to surface tension effects, a more conclusive study was needed to evaluate puncture site healing. The new test would not only definitively determine whether films had healed (both closed and sealed, i.e. airtight), but it would quantify the healing performance. As shown in Figure 3.10, the pressurized burst test (PBT) loaded the healed (or apparently healed) site with pressurized nitrogen gas at a steady rate until failure. Upon testing at RT, all four materials were shown to have healed with a nearly identical appearance (Figure 3.11). Surprisingly, analysis of the measured burst pressures also gave no indication of advantage provided by the ionic content. Results of this testing setup are given in Table 3.1. Here, the EMAA-0.3Na and EMAA-925 materials are shown to have performed favorably, healing in all repetitions with EMAA-925 being the most consistent. Though the ionic EMAA-0.3Na did hold greater pressure in at least half the cases, the thickness variation may have provided an effect. Additionally, comparison may reflect a disparity in strength of the specific material rather than the efficacy of the healing response because results are not normalized to the burst pressure for virgin (nondamaged) films. A more suitable test would have measured the virgin burst strength of the four films providing a measure of percentage healing. Although EMAA-960 was not consistent, the high ionic content EMAA-0.6Na performed the most poorly. Though samples appeared to close (successful stage I response), half of the films contained minimal or no pressure indicating a failure in the stage II component of healing.

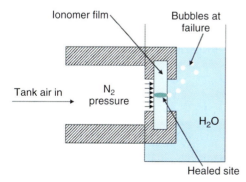

Fig. 3.10 Pressurized burst test (PBT) setup.

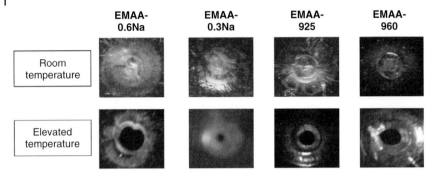

Fig. 3.11 Healing of EMAA films at room and elevated temperatures (reproduced from Kalista et al. [6], www.informaworld.com).

Table 3.1 Burst pressure of healed films following puncture at room temperature.

EMAA-0.6Na		EMAA-0.3Na		EMAA-925		EMAA-960	
Thickness (mm)	Burst pressure (MPa)	Thickness (mm)	Burst pressure (MPa)	Thickness (mm)	Burst pressure (MPa)	Thickness (mm)	Burst pressure (MPa)
0.91	<0.276	0.85	1.448	0.86	2.413	0.87	<0.276
0.93	0	0.86	1.482	0.86	2.482	0.88	1.448
0.97	<0.276	0.90	>3.034	0.92	2.413	0.88	1.999
0.99	1.790	0.94	>2.758	0.93	2.344	0.92	0
1.00	2.344	0.99	>3.103	0.94	2.620	0.92	1.931
1.04	2.275	0.99	2.689	0.97	2.482	0.93	2.413

Reproduced from Kalista *et al.* [6], www.informaworld.com.

The results of Kalista clearly show that ionic interaction was not the driving factor behind the healing response in EMAA materials. In fact, as observed in the higher ionic content samples, it may actually hinder adequate self-healing by decreasing the molecular mobility necessary for the melt interdiffusion stage of self-healing. The favorable performance of the low ionic content EMAA-0.3Na samples, however, suggests a benefit in the presence of ionic interactions. Further testing is necessary to make the determination, but it is suggested that there exists a balance between competing mechanisms of elastic recovery enhancement and the loss of melt interdiffusion provided by ionic interactions [6]. Therefore, the effect of ionic content on the healing ability must be reassessed.

3.3.5
Is Self-healing an Ionic Phenomenon? (Part II)

Though ionic content did not seem responsible for self-healing of films punctured at RT, it could play a significant role in healing at temperatures above and below RT.

Given the significant effect of temperature on polymers and especially on ionomers, which have a separate order–disorder transition, considerable differences might be expected. Kalista *et al.* reported the behavior of the same four materials punctured at elevated ($>60\,^\circ$C) [6] and subambient [5] temperatures.

As shown in peel tests of EMAA-0.3Na films, a test that mimics stage II sealing response, interfacial bond strength was shown to increase with increase in temperature with a significant rise at about $85\,^\circ$C [6]. Given the improved interdiffusion as bond temperature approached T_m, it was expected that a more complete ballistic self-healing response would be observed. However, for all tests of these films ($\sim$1 mm thickness), no healing was noted and a portion of the film was removed as shown in Figure 3.11. It was concluded that test temperature exceeded the temperature that is necessary for sufficient elastic recovery (stage I) to initiate closure and resulted in more global, permanent drawing. As shown in Figure 3.12, this "set" was more pronounced with increased temperature. At RT, the area around the puncture provided a rigid framework for the elastic storage and recovery of energy. However, as temperature increased, viscoelastic character shifted to a more viscous response of polymer at the puncture site and that of the material surrounding the hole. Rather than providing the needed framework for elastic retraction, the surrounding material absorbed and dissipated impact energy over a larger region preventing the mechanism required for stage I of self-healing.

Interestingly, a stage II process is suggested. For the unhealed samples at elevated temperatures, the tendency to form a circular hole (vs. that shown in Figure 3.8 for LDPE) under melt flow implies an attempt at healing by minimizing surface energy. This suggests that a melt interdiffusion (knitting) process is active, which, if it were to occur after hole closure, would have produced the necessary stage II sealing. This lends added support to the ballistic self-healing mechanism.

Given that the self-healing model suggests localized melting to be a key requirement for both stages I and II of the healing response, subambient testing was expected to prove equally unsuccessful at eliciting self-healing. As Fall [2] showed temperature to rise $\sim$5 $^\circ$C above the melt temperature for ballistic punctures at RT, the much larger temperature rise needed for puncture reversal below RT would be unexpected. Instead, the local temperature rise might remain in the submelt rubbery region producing a drawing/brittle fracture effect similar to LDPE at RT. However, Kalista and Ward [5] observed the healing response to persist for subambient temperatures of 10, -10, and $-30\,^\circ$C and the temperature rise was confirmed by DSC testing. To more accurately map the temperatures reached, additional study using the described DSC method or thermal IR would be necessary. Pressurized burst testing of the healed samples revealed similar performance to RT testing for the range of materials and the highly ionic EMAA-0.6Na remained the poorest performer. Interestingly, EMAA-925 was shown to reveal a significantly different behavior than the other three materials at $-30\,^\circ$C. The film fractured in a clearly brittle fashion with global cracking and the removal of a large portion of polymer as shown in Figure 3.13. It indicates that there was no attempt for any healing response. Although the temperature reached at the center of puncture is unknown, the polymer clearly experienced a change to glassy character. Localized melting

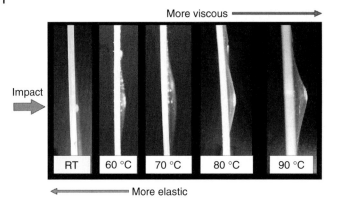

Fig. 3.12 Profile view showing permanent deformation in EMAA-0.3Na films with varying test temperature. Images obtained several hours after puncture (reproduced from Kalista et al. [6], www.informaworld.com).

may not have occurred or may have been over too small a region to allow projectile passage. Rather than elastic storage, decreased chain mobility surrounding the puncture site led to catastrophic film fracture. As testing did not go lower, the response cannot be correlated to other materials though a lower limit would certainly be expected. Additional testing by the described DSC method may help to better determine the temperatures reached.

3.3.6
Self-healing Stimulus

With the evidence presented above, the title of this chapter suggesting self-healing "ionomers" may be a misnomer. Indeed, the related nonionic EMAA copolymers also express the healing response. However, the term has already been widely used to discuss the ballistic self-healing phenomenon expressed by these materials. As a result there are many attempting to synthesize ionomer systems to self-heal ballistic and other types of damage, though none has been found successful in the literature.

While ionomers hold significant potential for a range of uses, they may or may not heal more effectively or over a wider range of test conditions than their nonionic counterparts (there is not enough information to determine thus far). However, ionic content does not seem to provide the specific stimulus for the self-healing behavior, at least in ballistic puncture. And, given the fact that increased ionic content in EMAA-0.6Na appears to hinder the melt interdiffusion process, which would likely be critical to healing damage in any thermally activated system, increased ionic content is not the answer. Although some ionic interaction may help, there clearly can be too much. This does not mean that ionic aggregates and the unique intermolecular attractions of the ionic architecture cannot provide a healing response; rather, it is not the cause in these self-healing systems. Therefore,

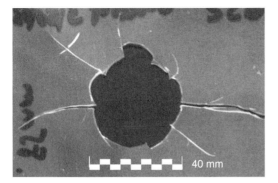

Fig. 3.13 EMAA-925 film following puncture at $-30°C$ indicating brittle fracture (adapted from Kalista and Ward [5]).

the pursuit and design of self-repairing systems based on this concept would wisely not limit itself to only the ionic domain.

Given the above summary, some conclusion or at least a hypothesis should be made as to the reason for self-healing in these materials. While ionic content is not the key answer, it is still possible that some intermolecular attraction or aggregating structure could contribute to producing the combination of thermomechanical properties necessary for the frictional heating, elastic retraction, and melt sealing necessary for self-healing. Fall and Kalista [2, 3, 5] both showed EMAA-925 and EMAA-960 to express a sub-T_m melt transition during DSC studies. Fall attributed it to a possible weak aggregation phenomenon given the lack of any indication in dynamic or viscoelastic experiments. In his work, Varley agreed with this assessment suggesting the responsible aggregating phenomenon may not be limited to an ionic one [32]. In fact, Bergman and Wudl [1] suggested that healing in the nonionic Nucrel may be due to its own physical cross-linking effect such that healing occurs via thermally controlled reversible hydrogen bonding, as shown schematically in Figure 3.14. This result is quite interesting and should be carefully considered. As EMAA ionomers have the same pendant acid structure in at least some of their chain architecture (EMAA-0.3Na is 30% neutralized or 70% nonneutralized), there would likely be some of the same interactions as those in nonionic materials. And if, as suggested by Varley [32], ionic clusters persist in the melt state during puncture, the hydrogen bonding phenomenon may be a more accessible reversible process. Therefore, intermolecular attractions beyond ionic may certainly play a role.

3.4
Other Ionomer Studies

In addition to ballistic testing, others have examined ionomer self-healing using modified techniques beyond ballistic puncture. Huber and Hinkley [28] designed an impression testing procedure to examine 6–10-mm-thick films of EMAA-0.3Na using a 1.6- or 0.8-mm-diameter flat-ended cylindrical probe. This test was proposed

as an alternative to projectile testing to assess new materials for the self-healing behavior. Examination was performed at rates of 0.5, 5, and 50 mm min^{-1} to a depth
of ~2.5 mm before removal a few seconds later. Given its much slower rate and
only partial depth of damage, the test is not a mimic of the ballistic test. However,
it allowed an instrumented approach to recording force as a function of impression
depth. Upon removal of the probe, the hole size in EMAA-0.3Na was shown to
have contracted smaller than the indenter (Figure 3.15). This happened for all three
speeds, though closure was most pronounced for the highest test rate as shown
in Figure 3.16. It might be suggested that probe removal in the direction opposite
puncture acted to collapse the impression; however, high-density polyethylene samples maintained the geometry and size of the probe tip. Additionally, if the probe tip
was left in the sample for additional time (overnight) rather than just a few seconds,
no healing was observed. Such a result is similar to the aforementioned nail test by
Kalista [3, 5], which inhibited the elastic closure response by the extended presence
of the nail allowing relaxation of the polymer chains rather than elastic recovery.

Huber and Hinkley repeated the impression test at 50 °C and 50 mm min^{-1}. In
comparison, the hole was shown to heal less effectively as shown in Figure 3.17.
Upon considering the nail test and other results above, this suggests that the

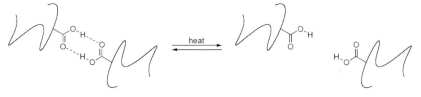

Fig. 3.14 Reversible hydrogen bonding in EMAA copolymers
[1] (Reproduced with permission by Springer Science and
Business Media).

Fig. 3.15 Profile view of partially healed EMAA-0.3Na film
tested using impression technique (reproduced from Huber
and Hinkley [28]).

increased temperature allowed a more rapid relaxation of the polymer chains. It may be especially true given this test to occur near or above the order/disorder temperature that would be expected to allow faster permanent set of the film rather than an elastic contraction on removal. Interestingly, the results might suggest a new form of healing in these materials. In their study, hole diameter is measured as a function of time at RT. The hole is shown to contract a few percent over four days of monitoring (Figure 3.17). This is consistent with the observation of Kalista *et al.* [6] for ballistic tests at elevated temperatures, in which nonhealed punctures minimized energy suggesting an interfacial knitting process (Section 3.3.4.). Additionally, Owen [27] observed a similar phenomenon for EMAA-0.3Na, though in a different damage mode. Films cut with a razor blade showed a gradual cut-healing process that blunted the cut, essentially "zipping up" the damage as shown in Figure 3.18. As would be expected, the extent of closure was observed to increase with temperature. The scenario is analogous to the interfacial healing process described in Section 3.3.1. It could be hypothesized that the contraction of the holes in Huber and Hinkley [28] was a similar event, only occurring much more slowly given the decreased polymer mobility below T_m as a gradual and progressive knitting of the hole perimeter over several days. Of course, it could also be a long-term relaxation, as Huber and Hinkley suggest, occurring by a recovery of anelastic strain having long relaxation times [28].

Huber and Hinkley [28] also applied the impression technique to a thermoplastic polyurethane elastomer (Pellethane 2102-65D, Dow Chemical), which had similar durometer hardness to EMAA-0.3Na. Though no size comparison is given, the hole

Fig. 3.16 Top view of partially healed EMAA-0.3Na samples following impression test for test rates of 50, 5, and 0.5 mm min^{-1} (left to right) at room temperature (reproduced from Huber and Hinkley [28]).

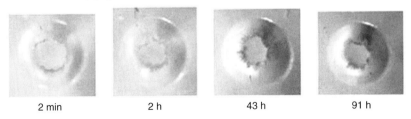

| 2 min | 2 h | 43 h | 91 h |

Fig. 3.17 Indented hole contraction with increased time (50°C) (reproduced from Huber and Hinkley [28]).

diameter was shown to decrease indicating a behavior more similar to the ionomer than the LDPE, suggesting a healing-like event. As a result, ballistic testing was also performed using a 9-mm pistol. Here, the elastomer film was reported to heal, though no picture or description of the extent of hole closure, healing, or sealing was given. Although a comparison to the ionomer ballistic healing cannot be made, the results indicate support for the conclusions for the healing stimulus in the EMAA materials. Varley [32] suggests cross-links provided by reversible hydrogen bonding in urethane systems may also play a role. However, its performance in stage II healing may be limited. This supports the idea where hydrogen bonding and favorable thermomechanical properties may combine to produce the healing response.

The most recent work by Varley and van der Zwaag [26] has proposed two new techniques for studying ionomer self-healing. These methods were designed to separately probe the elastic and viscous (stage I and II, respectively) components of the healing response using nonballistic methods. Specifically, EMAA-0.3Na films ($\sim$2 mm) were examined.

In the elastic test, a 9-mm-diameter disk was pulled through the film using a tensile testing machine as shown in Figure 3.19. Here, the disk was attached to a shaft that had already been partially pushed through the sample. Because impact speeds were much lower ($1-100$ m min^{-1}) and produce less frictional heating of the film, the disk was preheated before being pulled through and out of the film transferring the necessary heat. This allowed a controlled method for examining elastic response (stage I) at various temperatures and improved upon the work of Huber and Hinkley in mimicking projectile puncture as the impact motion occurs in the same constant direction with complete penetration and automatic shaft removal. A method for comparing percentage elastic healing for liquid flow through these puncture versus a 9-mm hole was also described. Results at 1 m min^{-1} showed an increasing elastic healing response as temperature increased with a plateau of 100%

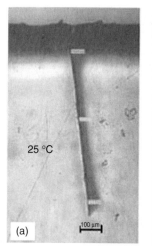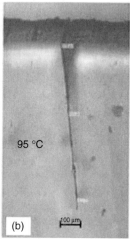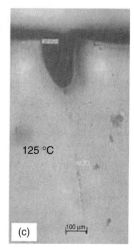

Fig. 3.18 Optical micrographs of cut healing with increase in temperature (reproduced from Owen [27]).

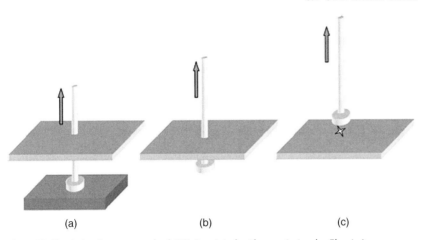

(a)   (b)   (c)

Fig. 3.19 Elastic healing test method [26] (reprinted with permission by Elsevier).

healing at 105 °C and above. Such a result is consistent with the self-healing model implying a melt condition necessary for elastic retraction of the impacted polymer zone in ballistic studies above. As shown in Figure 3.20, an increase in test speed from 1 to 100 m min^{-1} (at 110 °C) showed complete closure in all cases, though high speeds resulted in a smaller scar at the damage site. This is consistent with the results of Huber and Hinkley, whereby a quicker rate stores energy elastically rather than allowing the polymer to flow, relax, and conform to the damage shape. To provide a comparison to the ballistic method, SEM was used to show more similar features of the healed site for samples tested at 20 m min^{-1} versus those tested at 1 m min^{-1} indicating a shift to a better ballistic mimic test at higher speeds (for 110 °C). In additional work, Varley and van der Zwaag [41] have also examined the repeatability of elastic healing for samples at 1 m min^{-1} with complete recovery of the first three samples and a gradual decrease in ability thereafter (76% and 52% elastic healing after tests 3 and 6, respectively), as shown in Figure 3.21.

Varley and van der Zwaag [32, 41] have also examined the viscous (stage II, interfacial healing) response of the self-healing behavior in an isolated mimic experiment. Such a test is similar to the peel test performed by Kalista [3, 6], which examined interfacial healing below T_m. In the test by Varley and van der Zwaag, dog bone specimens were cut into two pieces and were placed in a tensile testing machine. They were heated while separate and brought into contact under pressure (~0.5 mm contracted from their initial total length) before cooling them rapidly. Tensile testing of these versus virgin control samples provided a measure of the percentage viscous healing efficiency. As shown in Figure 3.22, bond strength increased with temperature and bonding time (held at that temperature before cooling). These results are consistent with the expected increase in chain mobility and wetting at higher temperatures and the longer time for interdiffusion. The work by Fall [2] suggests that during ballistic puncture, the temperature rises only to ~98 °C, cooling to RT within 3 min. Therefore, the higher times and temperatures studied may not be a direct analog to the ballistic test; however, the healing efficiency

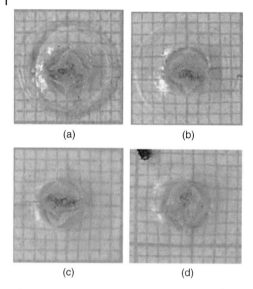

(a)　　　　　　　(b)

(c)　　　　　　　(d)

Fig. 3.20 Healed impact sites for (a) 1 m min^{-1}, (b) 5 m min^{-1}, (c) 20 m min^{-1}, and (d) 100 m min^{-1} for elastic testing at 110°C [26] (reprinted with permission by Elsevier).

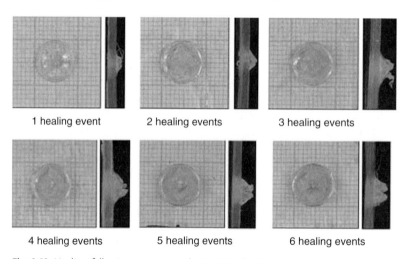

1 healing event　　　　2 healing events　　　　3 healing events

4 healing events　　　　5 healing events　　　　6 healing events

Fig. 3.21 Healing following successive elastic tests showing exit and profile images (reproduced from Varley and van der Zwaag [41]).

within the first 100 s would certainly prove quite relevant. Additionally, it is expected that ballistic punctures from different projectile types and speeds will result in a higher temperature at the damage zone. More significant thermal analysis of the ballistic puncture site must be performed to understand its effect on the viscous healing behavior. When the same experiment was repeated for linear low-density polyethylene (LLDPE) above T_m, testing proved difficult given its loss in structural rigidity in the melt [41]. Such an observation highlights the significant structural integrity provided by the ionic content in EMAA materials under melt conditions.

Although both the techniques are interesting, the elastic method is unique in its ability to examine the first stage of the ballistic self-healing process and represents an appreciable way to do so without using ballistic methods. Though the speeds were much slower than ballistic impact, which would generate a different viscoelastic response and difficulty in correlating to ballistic behavior, Varley and van der Zwaag utilized the interplay between rate and temperature well in their method. Though external heating is provided making the response a less "self-healing" one, the results indicate a promising and quite suitable method for examining the components of its behavior. With more comparative study between the elastic and ballistic techniques and examination of the heating profiles in both scenarios, more parallels can be drawn between the two and a more accurate mimic of ballistic testing will surely result. The research also provides interesting characterization of the self-healing behavior related to ionic content, a concept expanded on in a discussion of their work with composites below.

3.5
Self-healing Ionomer Composites

The potential for producing novel self-healing composites using EMAA ionomers has also been explored. Such systems would allow custom tailoring of ballistic self-healing systems with modified properties and function (e.g. containing conductive or magnetic fillers, a supplementary self-healing system to enhance or provide healing in cases where the response normally would not occur, etc.). Such exploration may also provide additional information on the self-repair response in EMAA materials.

In preliminary studies, Kalista successfully produced a novel, reinforced EMAA composite with the ability to self-heal upon projectile impact [3, 29]. Here, a melt-mixing process produced multiwall carbon nanotube(MWNT)-filled EMAA-0.3Na composites. SEM of fractured samples showed good dispersion of nanotubes in the polymer matrix (Figure 3.23) leading to mechanical property gains of ~15–20% in toughness, modulus, and tensile strength. Ballistic testing showed evidence of elastic closure as seen in Figure 3.24. These results compare favorably with that of EMAA-0.3Na shown in Figure 3.7, though a more prominent seam remains. However, no significant crater is produced. Although no pressure tests were performed on the healed site, the results provide viable proof of concept for the fabrication of reinforced or enhanced-function puncture-healing composites.

The development of healing EMAA composites was expanded in a later study. Here, Owen [27] developed an EMAA-0.3Na matrix composite system with the potential for a combined healing process. By design, this combined process would be able to self-heal not only ballistic puncture (though not tested), but it would also allow user-initiated healing of damage not repaired otherwise. Here, magnetic

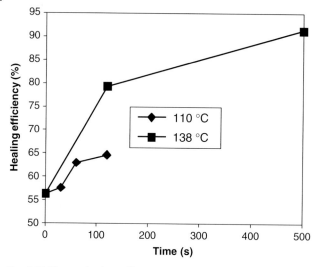

Fig. 3.22 Viscous healing efficiency as a function of time and temperature (reproduced from Varley and van der Zwaag [41]).

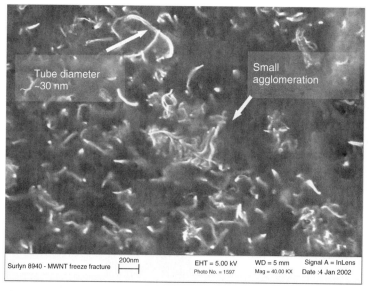

Fig. 3.23 SEM of fractured EMAA composite showing effective MWNT dispersion [3, 29].

nanoparticles were dispersed in an EMAA-0.3Na matrix using a melt pressing procedure. A magnetic inductive heating method was shown to successfully heat samples to temperatures that would self-heal cuts, though direct measures of cut healing using the inductive method were not obtained. Though not an autonomic technique, the study demonstrated the successful fabrication of ionomer systems

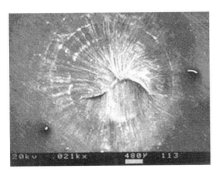

Fig. 3.24 SEM of healed EMAA composite (projectile entrance side) [3, 29].

with potential for an expanded healing capability. Future testing would be necessary to directly determine their performance as a magnetic inductive healing technique as well as in ballistic healing (though suggested possible in the composites of Kalista [3, 29]); however, the results present interesting potential.

Finally, Varley and van der Zwaag [41] have examined modified EMAA-0.3Na blends using a melt-dispersed zinc stearate ionic domain plasticizer in varying fractions (5, 10, and 20 wt%). This was proposed to explore the effect of changing ionic domain size, mobility, and strength on the projectile healing process. Ballistic puncture of these films yielded an appearance similar to that of EMAA-0.3Na, though SEM showed evidence of greater damage following puncture (larger cavities and other topographical features). Elastic healing response testing using the test previously described [26] yielded results similar to EMAA-0.3Na, though the repeatability of healing was significantly compromised. This was attributed to the mitigating effect of the zinc stearate plasticizer on elastic response. Indeed, increasing additive concentration led to reduced performance. Viscous response testing also yielded reduced performance in interfacial healing. This was justified by both the zinc stearate acting not only as a simple filler in reducing polymer diffusion, but also given its action as an ionic domain plasticizer. In sum, the results suggest an appreciable effect of ionic cluster properties on healing response in both ballistic puncture and the two testing schemes. It also represents further proof for modifying ionomer materials while maintaining the healing ability.

3.6
Conclusions

The self-healing behavior expressed by EMAA materials is unique and occurs as an autonomic process following ballistic puncture damage. Upon impact, the polymer is stretched locally both heating it to the melt state and storing energy elastically. The polymer then responds through a two-stage mechanism. In stage I, the molten polymer recovers elastically using the cool, surrounding polymer as a rigid framework. In stage II, this molten polymer is sealed through an interfacial healing process.

Both ballistic and nonballistic mimics have yielded much information about the healing response. With time, careful study of the ballistic test situation should allow nonballistic methods to more closely mimic the ballistic scenario allowing more controlled laboratory analysis. Ultimately, both methods will prove useful and are necessary in describing the self-healing capabilities. Finally, new and novel composite materials have been produced and are shown to also exhibit the self-healing response. Although only in the preliminary stages, they represent a meaningful start to tailoring of the design to scenarios where the virgin materials may not be suitable.

While the healing behavior was originally attributed to ionic interactions within the polymer during melt, ballistic testing indicated materials without ionic content also to heal. Surprisingly, ionic content was shown to provide a hindrance to healing at higher concentration. Therefore, it should be noted that self-healing in these materials is not specifically an ionic (or ionomer) phenomenon. Instead, it was proposed that it may actually be thermally reversible physical cross-links and intermolecular interactions, rather than the aggregating behavior provided by ionic interactions, which produce the thermomechanical properties responsible for the self-healing response. As such interactions should exist for both the ionic and nonionic copolymers, this hypothesis seems reasonable.

In any case, EMAA ballistic self-healing remains an interesting phenomenon with significant potential for application given its ability to heal under such unique conditions and at such a rapid rate. As more exposure develops, the added focus on this unique system should yield significant advances in determining the exact cause of healing and in designing and synthesizing new polymeric materials possessing similar ability.

Acknowledgements

The author would like to specifically acknowledge the support of the Department of Physics and Engineering at Washington and Lee University, Russell Varley at CSIRO, Tom Ward at Virginia Tech and Mia Siochi at NASA Langley Research Center. Financial support from NASA (NAG-1-03058), the NSF IGERT Program at Virginia Tech, and the donation of materials from DuPont is gratefully acknowledged. The author would also like to recognize the organizers and attendees of the First International Conference on Self Healing Materials and particularly those at the Delft Centre for Materials (DCMat) at TUDelft for making meaningful conversations and collaborations possible in this new and emerging field.

References

1 Bergman, S. D. and Wudl, F. (2007) Re-Mendable Polymers, in *Self Healing Materials: An Alternative Approach to 20 Centuries of Materials Science* (ed. S. van der Zwaag), Springer, Dordrecht.

2 Fall, R. (2001) *Puncture Reversal of Ethylene Ionomers — Mechanistic Studies*, Master of Science, Virginia Tech, Blacksburg.

3 Kalista, S. J. (2003) *Self-Healing of Thermoplastic Poly(Ethylene-co-*

Methacrylic Acid) Copolymers Following Projectile Puncture, Master of Science, Virginia Tech, Blacksburg.

4 Kalista, S. J. and Ward, T. C. (**2003**) Proceedings of the 26th Annual Meeting of the Adhesion Society, pp. 176–78.

5 Kalista, S. J. and Ward, T. C. (**2007**) *Journal of the Royal Society Interface*, **4**(13), 405–11.

6 Kalista, S. J., Ward, T. C. and Oyetunji, Z. (**2007**) *Mechanics of Advanced Materials and Structures*, **14**(5), 391–97; www.informaworld.com (access 2008).

7 Kalista, S. J. and Ward, T. C. (**2004**) Proceedings of the 27th Annual Meeting of the Adhesion Society, Inc., pp. 212–14.

8 Kalista, S. J. and Ward, T. C. (**2007**) Self-Healing of Poly(Ethylene-co-Methacrylic Acid) Copolymers Following Ballistic Puncture, *in Proceedings of the First International Conference on Self Healing Materials*, Springer, Noorwijk aan Zee.

9 White, S. R. *et al.* (**2001**) *Nature*, **409**(6822), 794–97.

10 Pang, J. W. C. and Bond, I. P. (**2005**) *Composites Science and Technology*, **65**(11–12), 1791–99.

11 Pang, J. W. C. and Bond, I. P. (**2005**) *Composites Part A: Applied Science and Manufacturing*, **36**(2), 183–88.

12 Rees, R. W. and Vaughan, D. J. (**1965**) *Polymer Preprints, American Chemical Society, Division of Polymer Chemistry*, **6**, 287–95.

13 Tant, M. R. and Wilkes, G. L. (**1987**) Viscoelastic behavior of ionomers in bulk and solution, in *Structure and Properties of Ionomers* (eds M. Pineri and A. Eisenberg), D. Reidel Publishing, Dordrecht, pp. 191–226.

14 Eisenberg, A. and Rinaudo, M. (**1990**) *Polymer Bulletin*, **24**(6), 671.

15 Eisenberg, A. and Kim, J.-S. (**1988**) *Introduction to Ionomers*, Wiley-VCH Verlag GmbH, New York.

16 Schlick, S. (ed.) (**1996**) *Ionomers: Characterization, Theory, and Applications*, CRC Press, Boca Raton.

17 Eisenberg, A. (**1970**) *Macromolecules*, **3**(2), 147–54.

18 Eisenberg, A., Hird, B. and Moore, R. B. (**1990**) *Macromolecules*, **23**(18), 4098–107.

19 Bellinger, M., Sauer, J. A. and Hara, M. (**1994**) *Macromolecules*, **27**(6), 1407–12.

20 Statz, R. (**1986**) in *Ethylene Copolymer Ionomers, in History of Polyolefins: The World's Most Widely Used Polymers* (eds R. B. Seymour and T. Cheng), D. Reidel Publishing, Dordrecht, pp. 177–92.

21 Rees, R. (**1987**) Ionomeric thermoplastic elastomers early research–surlyn and related polymers, in *Thermoplastic Elastomers: A Comprehensive Review* (eds N. R. Legge, G. Holden and H. E. Schroeder), Carl Hanser Verlag, Munich, pp. 231–43.

22 Hara, M. and Sauer, J. A. (**1994**) *Journal of Macromolecular Science: Reviews in Macromolecular Chemistry and Physics*, **34**(3), 325–73.

23 Tadano, K. *et al.* (**1989**) *Macromolecules*, **22**(1), 226–33.

24 http://www2.dupont.com/Nucrel/en__US/.

25 http://www2.dupont.com/Surlyn/en__US/.

26 Varley, R. J. and van der Zwaag, S. (**2008**) Development of a quasi-static test method to investigate the origin of self-healing in ionomers under ballistic conditions. *Polymer Testing*, **27**(1), 11–19.

27 Owen, C. C. (**2006**) *Magnetic Induction for In-situ Healing of Polymeric Material*, Master of Science, Virginia Tech, Blacksburg.

28 Huber, A. and Hinkley, J. (**2005**) *Impression Testing of Self-Healing Polymers*, NASA Technical Memorandum, NASA/TM-2005-213532.

29 Kalista, S. J. and Ward, T. C. (**2006**) Self-Healing in Carbon Nanotube Filled Thermoplastic Poly(Ethylene-Co-Methacrylic Acid) Ionomer Composites. *Proceedings of the 29th Annual Meeting of the Adhesion Society, Inc.*

30 Varley, R. J. and van der Zwaag, S. (**2007**) An investigation of the self healing mechanism of

ionomer based systems, *Proceedings of the First International Conference on Self Healing Materials*, Springer, Noordwijk aan Zee.

31 Coughlin, C. S., Martinelli, A. A. and Boswell, R. F. (2004) *Mechanism of Ballistic Self-Healing in EMAA Ionomers.* Abstracts of Papers of the American Chemical Society, 228, p. 261.

32 Varley, R. J. (2007) Ionomers as self healing polymers, in *Self Healing Materials: An Alternative Approach to 20 Centuries of Materials Science* (ed. S. van der Zwaag), Springer, Dordrecht.

33 Seibert, G. M. (1996) Shooting range targets. U.S. Patent. #5,486,425, Jan. 23, 1996.

34 Wool, R. P. (1995) *Polymer Interfaces: Structure and Strength*, Hanser/Gardner Publications, Cincinnati.

35 Kim, Y. H. and Wool, R. P. (1983) *Macromolecules*, **16**(7), 1115–20.

36 de Gennes, P. G. (1971) *Journal of Chemical Physics*, **55**(2), 572–79.

37 Boiko, Y. M. *et al.* (2001) *Polymer*, **42**(21), 8695–702.

38 White, S. R. (2007) *The Future of Autonomic Materials Systems.* First International Conference on Self Healing Materials, Noordwijk aan Zee.

39 Siochi, E. J. and Bernd, J. NASA Langley Research Center, unpublished results/private communication.

40 Kalista, S. J. unpublished work.

41 Varley, R. J. and van der Zwaag, S. (2008) Acta Materialia (*in press*).

42 Kalista, S. J. and Sivertson, E. (2006), unpublished work.

4

Self-healing Anticorrosion Coatings

Mikhail Zheludkevich

4.1
Introduction

The huge economic impact of corrosion of metallic structures is a very important issue all over the world. Developed countries lose about 3% of national product every year due to corrosion degradation. The corrosion impacts our daily life causing, very often, not only the economic consequences but also life-threatening situations. Engineering structures that can be significantly damaged and eventually destroyed by corrosion include bridges, pipelines, storage tanks, automobiles, airplanes, and ships. The most common environments causing corrosion are atmospheric moisture, natural waters, and man-made solutions.

Application of organic coatings is the most common and cost-effective method of improving the corrosion protection and, thereby, the durability of metallic structures. A wide range of engineering structures, from cars to aircrafts, from chemical factories to household equipment, is effectively protected by the coating systems. The main role of an organic polymer coating in corrosion protection is to provide a dense barrier against corrosive species. Along with these barrier properties, resistance to a flow of charge, electronic and ionic, is also important since the corrosion processes have electrochemical nature and the transfer of charge is involved. However, defects appear in the organic protective coatings during exploitation of the coated structures opening a direct access for corrosive agents to the metallic surface. Also, exposure of these coatings to aqueous electrolyte solutions causes many coatings to swell and develop conductive pathways not present in the coating before the exposure. The corrosion processes develop faster after disruption of the protective barrier. Therefore, an active "self-healing" of defects in coatings is necessary in order to provide long-term protection effect.

The term *self-healing* in materials science means self-recovery of the initial properties of the material after destructive actions of external environment. The same definition can be applied to functional coatings. However, a partial recovery of the main functionality of the material can be also considered as a self-healing ability. Thus, in the case of corrosion protective coatings the term *self-healing* can

Self-healing Materials: Fundamentals, Design Strategies, and Applications. Edited by Swapan Kumar Ghosh
Copyright © 2009 WILEY-VCH Verlag GmbH & Co. KGaA, Weinheim
ISBN: 978-3-527-31829-2

be interpreted in different ways [1, 2]. The classical understanding of self-healing is based on the complete recovery of the coatings' functionality due to a real healing of the defect. However, the main function of anticorrosion coatings is the protection of an underlaid metallic substrate against an environment-induced corrosion attack. Thus, it is not obligatory to recuperate all the properties of the film in this case. The hindering of the corrosion activity in the defect by the coating itself employing any mechanisms can be already considered as self-healing, because the corrosion protective system recovers its main function, namely, the corrosion protection, after being damaged. Both self-healing concepts will be reviewed in the present work.

The protective coating system usually consists of several functional layers [3, 4]:

- a conversion coating or a pretreatment that provides good adhesion between metal and an organic polymer layer and eventually confers an additional active corrosion protection;
- an inhibition primer that contains a leachable anticorrosive pigment for effective corrosion inhibition;
- a topcoat composed of a robust polymer layer aiming in the required appearance performance and a good physical barrier against active corrosive agents.

Most effective active corrosion protection systems for metals to date contain chromates. Hexavalent chromium-containing compounds are a traditional type of corrosion inhibitor that exhibits far better corrosion resistance than any other inhibitors due to the strong oxidation properties of Cr^{6+} [5–7]. When chromate-based coatings are applied, the hexavalent chromium compounds, loosely bonded in the film, are slowly leached out during exposure to aqueous media healing scratches and other defects [8, 9]. Chromate anions inhibit the corrosion being released from the protection system in the defected area, migrating then, to the metal surface and reacting with actively corroding sites, thereby resulting in passivation [10]. Chromate can interact with a metal surface forming a $Cr_2O_3 \cdot n\text{-}H_2O$ solid hydrated conversion coating. Chromate acts both as cathodic and anodic inhibitor reducing the overall electrochemical activity of the surface [5]. Even very low concentration of chromate is enough to hinder the corrosion activity of metallic structures. For example, 10^{-5} M CrO_4^{2-} can effectively suppress cathodic reaction of oxygen reduction on AA2024 [11]. A significant decrease in metastable pit formation was achieved by Pride et al. [12] using as little as $2.5 \cdot 10^{-5}$ M CrO_4^{2-}, while a four times higher concentration is sufficient to reduce a stable pit initiation on unpassivated panels of AA2024-T3 [13].

Usually, conversion coatings on metals also contain chromates. The electrochemical route for preparation of a conversion coating on metal via anodization process very often employs chromic acid anodizing baths [7]. Although chromate has been shown to leach from the conversion coatings, the most part of the chromate, needed for the protection of the surface, is released from the primer due to the low loading capacity of the thin conversion coating layer [8].

Chromate-inhibited primers are widely used as a component in self-healing corrosion protection systems for a range of metallic substrates. Chromate-based pigments, such as $BaCrO_4$ or $SrCrO_4$, are typically introduced to an epoxy polyamide matrix, together with other functional pigments [8, 9]. The chromate pigments $BaCrO_4$ and $SrCrO_4$ are selected for their low, but finite, solubility, their efficient inhibition against corrosion for a range of metals, and compatibility with the polymer matrix [10]. The solubility of the chromate pigment is a crucial issue from the standpoint of corrosion protection. Low solubility results in negligible inhibiting power while high solubility causes extensive osmotic blistering effect. A protection system with chromate-loaded primer, therefore, not only provides a barrier to the aggressive corrosive species, but also contains a reservoir of inhibitors that can heal the defects in the coating [14]. Release of chromate from the primer is a complex process, which involves a number of stages: transport of water into the defect and across the damaged edge of the primer; transport of water into the primer to the pigment particles; dissolution of the inhibitor; and transport of the inhibitor through the primer to the defect site [8].

Despite the good corrosion protection properties, the leachability and superior oxidation properties of the chromates make them environmentally unfriendly. Skin contact, inhalation, and ingestion can cause penetration of the chromium waste into the human body. The hexavalent chromium species can be responsible for DNA damage and cancer [7]. Therefore, environmental regulations will ban the use of Cr^{6+}-containing compounds in the corrosion protection systems in Europe from 2007. The development of nonchromate environmentally friendly active corrosion protective systems is, therefore, an issue of prime scientific and technological importance for various industries due to the formed gap between industrial needs and currently existing corrosion protection technologies.

The recent development in the area of new environmentally friendly self-healing anticorrosion coatings is shortly overviewed here. The first part of this review is focused on the coatings that can heal the defects recuperating the coating integrity or sealing the defects mechanically by corrosion products. And the second part of the paper is devoted to the coatings that can provide active corrosion protection of metal in the defected areas by electrochemical and/or inhibition mechanisms.

4.2
Reflow-based and Self-sealing Coatings

4.2.1
Self-healing Bulk Composites

The continuous improvement of performance of man-made materials in many engineering structures is a key issue in determining the reliability and cost-effectiveness of such systems. One of the main focuses of current research efforts is the development of new bioinspired self-repairing materials. Many natural *materials* are themselves self-healing composites [15]. The repair strategies of living organisms attract the materials designers looking for low-weight structures with enhanced

service life. These bioinspired approaches do not completely imitate the real biological processes involved because the latter are too complex. Instead, the designers
of self-healing materials try to combine traditional engineering approaches with
biological self-healing mechanisms.

Several self-healing polymer composites were recently reported. An outstanding
example is an epoxy-based system able to heal cracks autonomically as described by
White *et al.* [16]. The polymer bulk material contains a microencapsulated healing
agent that is released upon crack initiation. Then, the healing agent is polymerized
after the contact with the embedded catalyst bonding the crack faces and recovering the material toughness. Dicyclopentadiene-filled microcapsules (50–200 μm)
with a urea-formaldehyde shell were prepared using standard microencapsulation
techniques. The polymer shell of the capsule provides a protective barrier between
the catalyst and the monomer to prevent polymerization during the preparation of
the composite. However, the chemistry of this system had a significant drawback
due to the probability of side reactions with the polymer matrix and air. A new
advanced self-healing polymer system based on tin-catalyzed polycondensation
of phase-separated droplets containing hydroxyl end-functionalized polydimethylsiloxane and polydiethoxysiloxane was recently suggested by the same group [17].
The concept of phase separation of the healing agent was employed in this work
simplifying the processing greatly as the healing agent was just mixed into the
polymer matrix. The catalyst, di-*n*-butyltin dilaurate (DBTL) microencapsulated
within polyurethane shell, which was embedded in a vinyl ester matrix is released
when the capsules are broken by a mechanical damage as shown in Figure 4.1.
The released catalyst initiates polymerization of monomers from the surrounding
droplets healing the propagating crack. This new system confers high stability at
humid environments and high temperatures ($>100\,^{\circ}$C).

The works of White were rapidly followed up by other groups and many
other self-healing systems were developed, such as self-validating adhesives [18]
and self-healing epoxy composites [19, 20]. The original idea of hollow spheres
was extended for hollow reinforcement fibers, used in fiber reinforced plastic,
embedding a liquid resin. The repair process in this case is triggered after an
impact loading of the material [15, 21–23]. The hollow glass fibers ranging in
diameter from 30 to 100 μm and hollowness up to 65% can be filled with uncured
resin systems that bleed into a damage site upon fiber fracture, as shown in
Figure 4.2. After being cured they provide a method of crack blocking and recovery
of mechanical integrity.

Even more advanced approach was recently suggested employing a self-healing
system capable of autonomous repairing of repeated damage events [24]. The
substrate composite delivers the healing agent to the cracks in a polymer coating
via a three-dimensional microvascular network embedded in the substrate. Crack
damage in the coating is repaired repeatedly mimicking a body/skin system.
However, this approach cannot be used in the case of corrosion protective coatings
since a microvascular network cannot be created in a metallic substrate.

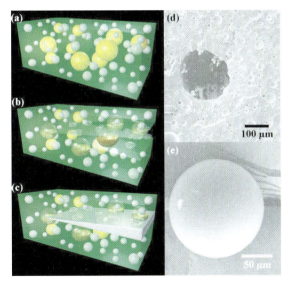

Fig. 4.1 Schematic of self-healing process: (a) self-healing composite consisting of microencapsulated catalyst (yellow) and phase-separated healing agent droplets (white) dispersed in a matrix (green); (b) crack propagating into the matrix releasing catalyst and healing agent into the crack plane; (c) a crack healed by polymerized polydimethylsiloxane (crack width exaggerated). Scanning electron microscopy (SEM) images of (d) the fracture surface, showing an empty microcapsule and voids left by the phase-separated healing agent and (e) a representative microcapsule showing its smooth, uniform surface. Reprinted from [17] with permission from Wiley-VCH Verlag GmbH & Co. KGaA.

4.2.2
Coatings with Self-healing Ability based on the Reflow Effect

Transfer of the self-healing approaches used for bulk materials to the coatings is very complicated since the self-healing system should be embedded in a thin submillimeter polymer layer. The idea of a coating suitable for the reflow-healing of defects was patented for the first time almost a decade ago [25]. The invention describes protective coatings with *in situ* self-repair ability after being damaged by careless handling or stressful environments. The repair is achieved through microencapsulated polymerizable agents incorporated into the coating matrix, as in the case of bulk materials described above. If the coating is damaged, the microcapsules rupture leaching the film forming components in the immediate vicinity of the damage. The fluid flows over exposed areas of the metal surface and fills any defects or cracks in the coating renewing the protective barrier. One of the first works in this area was focused on incorporation of sealant-containing vesicles that release the healing agent under destructive mechanical or chemical impact. The addition of the microencapsulated polymerizable agent to the fusion bonded epoxy coating was discussed as an advanced approach to design a more

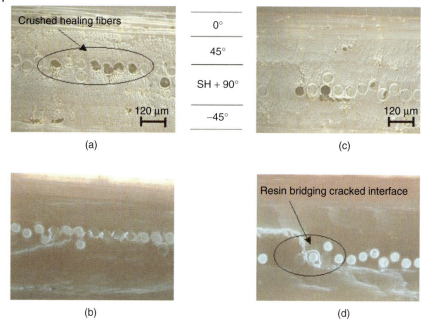

Fig. 4.2 Crushed-healing fibers located under the impact site viewed under normal (a) and UV (b) illumination. Healing resin bridging cracked interface viewed under normal (c) and UV (d) illumination. Reprinted from [22] with permission from the Royal Society.

damage resistant film than the traditional ones [26]. The mechanical impact, which damages the conventional coatings, leads to a release of the active agent from the microcapsules, thereby healing the protective coating.

Kumar *et al.* attempted to introduce different types of capsules loaded with corrosion inhibitors and coating repair compounds into commercially available paints [27]. The efficacy of "self-healing" corrosion protection coatings with urea-formaldehyde and gelatin microcapsules (50–150 μm diameter) containing several types of film forming agents has been studied. The microcapsules stay intact for a long time in the dry coatings, as shown in Figure 4.3a, and are ruptured only by a damage releasing the core constituents to the defect. The chosen microcapsules are stable for more than two weeks even in paint formulations. However, the results of the experiments suggest that they should be mixed with paint preferably at the time of application. Moreover, the best results were obtained when the microcapsules were sprinkled as a discrete layer on top of a thin layer of previously applied primer and, then a second layer of primer was deposited followed by a topcoat layer (see Figure 4.3b). The corrosion protection performance in this case was shown to be far superior compared to the mixing of the microcapsules in the primer before being applied. Accelerated corrosion tests of these experimental coatings based on ASTM D 5894 indicate that incorporation of the self-healing

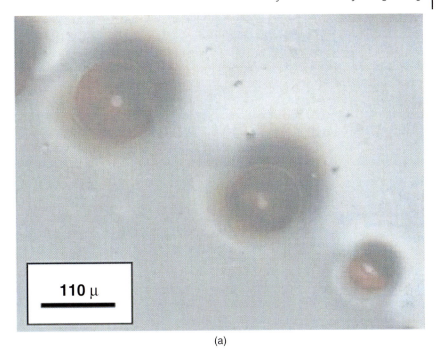

(a)

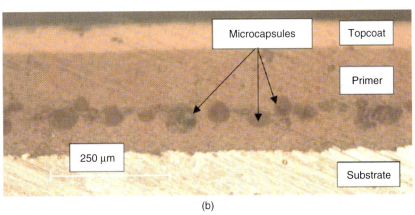

(b)

Fig. 4.3 (a) Micrograph of gelatin capsules in polyurethane paint after 2 h; (b) optical micrograph of cross section of coating with microcapsules. Reprinted from [27] with permission from Elsevier.

microcapsules into commercially available primers can remarkably reduce under film corrosion on steel enclosures for outdoor equipment.

Sauvant-Moynot *et al.* suggested using a self-healing coating together with a cathodic protection system. Specific film formers sensitive to pH and electrical field were introduced to the coatings applied on metal structures and were used under cathodic protection [28]. A dried water soluble and self-curable epoxy electrode-positable adducts as fillers (30 wt%) were used as organic film formers. A significant reduction in the current needed for cathodic protection was revealed demonstrating the self-healing ability of the coatings under study. The barrier properties were significantly increased in comparison to scratched reference samples.

The idea of reflow-healing of protective coatings has already found its commercial realization. Nissan has recently announced the "Scratch Guard Coat" painting system, which contains a newly developed high-elastic resin providing reflow in artificial scratches [29]. The new coating system is effective for about three years and is five times more resistant to abrasions caused by a car-washing machine compared with a conventional clear paint.

4.2.3
Self-sealing Protective Coatings

Examples of self-healing coatings presented above are based on the polymerization of a healing agent in the defects recovering the barrier properties of the protective coatings. However, the barrier properties of a damaged coating can also be recuperated by a simple blocking of the defects with insoluble precipitates. These deposits in cracks can originate from reaction of corrosive medium or corrosion products with components of self-healing coating, though the term *self-sealing* seems to be more appropriate in this case.

Sugama *et al.* developed a poly(phenylenesulfide) (PPS) self-sealing coating for carbon steel heat exchanger tubes, used in geothermal binary-cycle power plants operating at brine temperatures up to 200 °C [30]. Hydraulic calcium aluminate (CA) fillers containing monocalcium aluminate ($CaO \cdot Al_2O_3$) and calcium bialuminate ($CaO \cdot 2Al_2O_3$) reactants as major phases were dispersed in the coating matrix. The decalcification–hydration reactions of the $CaO \cdot Al_2O_3$ and $CaO \cdot 2Al_2O_3$ fillers, surrounding the defect, lead to a fast growth of boehmite crystals in the cracks. The block-like boehmite crystals (~4 μm in size) fill the cracks after 24 h effectively sealing them, as demonstrated in Figure 4.4. The sealing of the scratch causes an increase, about 2 orders of magnitude, in the pore resistance of the coating suggesting that the conductive pathway for aggressive species is thoroughly blocked. Extension of the exposure time to 20 days results in a stable value of pore resistance meaning that the sealing of the cracks by boehmite crystals plays an essential role in the recovery of the protective function of a PPS coating.

Another self-sealing approach was used by Hikasa *et al.* for creating a self-repair ceramic composite protective coating [31]. Sodium–clay (hectorite) and silica multilayers were deposited using spin coating technique. The hectorite is swelling clay that expands cubically due to the reaction with water. When water penetrates into the crack in clay/silica composite film it reacts with hectorite causing its swelling and consequent blocking of the defect.

The encapsulation strategy was also suggested to prepare a polymeric self-sealing coating [28]. Epoxy-amine microcapsules containing $MgSO_4$ and ranging in size from 10 to 240-μm diameter were prepared by interfacial polymerization in inverse

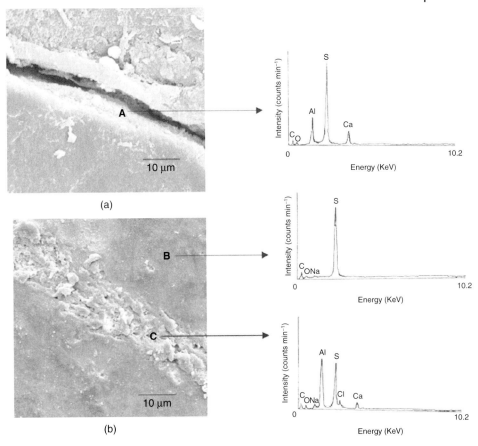

Fig. 4.4 SEM images coupled with Energy Dispersive X-ray Analysis (EDX) spectra for cleaved PPS coatings with 5 wt% CA fillers before (a) and after (b) exposure for 24 h to CO_2-laden brine at 200 °C. Reprinted from [30] with permission from Elsevier.

emulsion. A commercial liquid epoxy-amine paint doped with microcapsules and applied to a steel substrate was tested under cathodic protection conditions. Magnesium sulfate was chosen as a healing agent since it can form insoluble $Mg(OH)_2$ deposits at high pH, which can arise in paint defects due to cathodic processes. The idea of this work was to achieve the sealing of the defects by the hydroxide formed from magnesium ions released from the capsules. But the authors did not succeed with this and the self-sealing was not achieved in this case.

4.3
Self-healing Coating-based Active Corrosion Protection

The examples reviewed above demonstrate the coatings that are able to heal or seal the defects blocking them mechanically via polymerization or precipitation mechanisms. The physical integrity of the coating is partially recovered due to these processes. However, the main function of the protective coatings is to protect the underlaid substrate from corrosion processes. The partial blocking of the pathways for corrosive species very often does not mean an effective hindering of the corrosion. Corrosion processes can occur even under the coating in place of the healed or sealed defect if the electrolyte has already penetrated into the substrate when the defect was induced. Therefore, very often, other strategies are used to achieve an active suppression of the corrosion processes in the defected areas. The coating systems, which use active corrosion protection mechanisms, can also be considered as "self-healing" coatings since the main function of the protection system, namely, the corrosion protection, is recovered during operation. A short overview of different active corrosion protection strategies and self-healing coatings based on them is presented in the following sections.

4.3.1
Conductive Polymer Coatings

Electroactive conductive polymer (CPs) are currently being explored for use in corrosion control coating systems as possible environmental-friendly alternatives for the Cr(VI)-based coatings. The electroactivity and the electronic conductivity of CPs set them apart from traditional organic coatings [32]. Conjugated polymers that exhibit electroactivity and/or display some level of conductivity or semiconductivity are considered as CPs. They are partially oxidized (p-doped) and, thus, contain counteranions for overall charge neutrality.

First observations of corrosion protection of steel by polyaniline were reported in 1981 by Mengoli et al. [33]. This work was rapidly followed up by other groups and a number of papers have been reported on the protective properties of CPs against corrosion of iron [34], mild steel [35, 36], stainless steel [37], titanium [38], copper [39], zinc [40], and aluminum alloys [41, 42]. Hundreds of works devoted to the anticorrosive properties of CPs exist today in the literature, which is well overviewed in a number of thematic reviews [32, 43–48]. Here, main attention is paid only to the general principles of active corrosion protection conferred by CPs and to the description of main types of CPs.

Most CPs used in corrosion protection can be categorized into the following three classes: polyanilines, polyheterocycles, and poly(phenylene vinylenes). CPs can be synthesized using chemical or electrochemical routes. Most CPs can be anodically oxidized producing a conducting film directly on the surface [43].

Such polymers can also be classified based on the nature of the p-doping process during the polymer synthesis [32]:

1. *Type 1:* protonic/electronic doping involving both proton and anion incorporation into the polymer (e.g. polyaniline).
2. *Type 2:* electronic doping with anion incorporation (e.g. polypyrrole or polythiophene).
3. *Type 3:* electronic doping with cation expulsion, either from a covalently attached acid group (also called *self-doped*, e.g.

sulfonated polyaniline) or from a sufficiently large, physically
entrapped, immobile acid (or salt, e.g. a polyelectrolyte).

All three types of polymers can be used directly as a primer coating, as a surface
treatment, or as a component blended with more conventional coatings.

The main mechanism of corrosion protection by CPs originates from their
electroactivity and electronic conductivity since the CPs are redox-active materials
with equilibrium potentials that are noble relative to that of iron and aluminum
and similar to the potential of chromate. The CPs anodically polarize the metal in
the presence of an electrolyte. The polarization forces the anodic oxidation of the
metal shifting its potential to the passive zone. This can be achieved for metals
which are able to form a passsive oxidelhydroxide film on the surface. This is the
main mechanism of a so-called anodic protection of metals. Thus, a CP coating can
shift the metal potential into the passivity region leading to a formation of a passive
film, consequently drastically decreasing the corrosion current. The CP film plays
a role of a cathode where the oxygen reduction process takes place. This active
protection mechanism is shown in Figure 4.5. Passivation of the defect caused by
the action of the coating system leads to the recovery of the main function of the
protective coating, which can be considered as self-healing.

The anodic protection approach implies that the metal at some conditions can be
in the passive state. However, for many metals in some specific environments, for

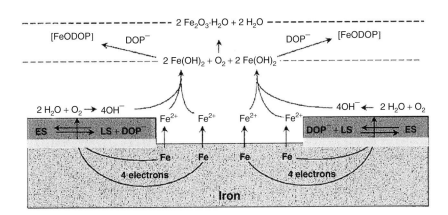

Fig. 4.5 Mechanistic view of pinhole passivation by conductive polymer coating. Reprinted from [49] with permission from the Electrochemical Society.

example, for aluminum alloys in neutral NaCl solutions, the passivity state cannot be achieved. In this case, other factors such as metal complexation and counterion release, where the counterion might be selected for its known corrosion inhibition, may also be important [49, 50].

The counterion of CP can play a crucial role in active corrosion protection, especially when emeraldine salt primers are used. For example, the corrosion protection increases in the following order: polymethoxyaniline sulfonic acid < tartaric acid < dinonylnapthalene sulfonic acid < dodecylbenzene sulfonic acid [51]. The phosphonic acid dopants exhibit even better anticorrosion effect when compared to that of sulfonic ones [49]. These examples demonstrate that the active corrosion protection effect is caused not only by the anodic protection, originating from the electroactivity of the CP film, but also by the inhibiting action of the counterion (dopant). In these cases, the CP coatings can offer an active anticorrosion effect even for metallic substrates that cannot be passivated at certain environmental conditions [50]. Kendig *et al.* developed "smart" corrosion protection CP coatings doped with different corrosion inhibitors [52]. It was found that the corrosion reaction in defect area on aluminum alloy 2024 appears substantially slowed with the release of the corrosion inhibitor from the CP film. Moreover, the conductive coating behaves in this case as a "smart" reservoir of the corrosion inhibitor. The release of inhibiting ions is triggered by electrochemical activity in the defect and occurs only on demand when the corrosion process is started.

Currently, a lot of attention is paid to the application of CPs as dispersions in conventional paint systems [48].

CP coatings and CP-based pigments have been already investigated from the standpoint of corrosion protection for more than 25 years. However, there are not many real industrial applications of these systems reported up to the present date. This seems to appear because the coatings used for industrial applications must meet many requirements such as high chemical stability when comes in contact with air and water, good compatibility with conventional organic paint systems, good adhesion on the metal and on the topcoats, high electrical conductivity, and high electroactivity. None of the CP systems, known up to now, has all these properties, which limits the real industrial application of the CPs in the self-healing anticorrosion coatings.

Another approach for active corrosion protection and self-healing in corroded defects is the use of chemical inhibiting species which can be released from the coating system hindering the corrosion activity. The corrosion inhibitor can be added to the different parts of the coating system since the corrosion protection coating is usually a complex system constituted by several layers. The inhibiting compounds can be incorporated into the conversion coating, the primer, or the top-coat using different strategies. The layer containing the corrosion inhibitor serves, then, as a reservoir leaching the inhibitor out during its service life. The following sections of this chapter overviews different strategies of the inhibitor incorporation into the corrosion protective system in order to achieve the self-healing effect and active corrosion protection.

4.3.2
Active Anticorrosion Conversion Coatings

The coating adjacent to the metal layer of a protective system plays one of the most important roles, namely, it provides the adhesion between the metal surface and the polymer layer. Chemical pretreatments or conversion coatings are mainly employed to serve this purpose. Chemical conversion coatings are usually inorganic films formed directly on the metal surface due to its interaction with the components of the pretreatment solution [53]. Such coatings, formed as an integral part of the metal surface, are inorganic crystalline or amorphous surface films that are much less reactive to subsequent corrosion than the original metal surface. This film neutralizes the potential of the local anodic and cathodic galvanic corrosion sites leading to an equal potential on the metal surface. Conversion coatings are also absorptive bases for improving the adhesion to paints. Chemical conversion coatings have a high adherent character and provide high speed of coating application, which makes them preferable compared to electrochemical surface modification methods. They are economical and can be formed without an application of any external potential using simple equipment. Historically there are three main chemical conversion coating processes, which are classified as phosphating, chromating, and oxalating.

The oxalating and phosphating conversion coatings improve the passivation of metal and enhance adhesion between metal and organic coating. The chromate coatings provide additionally an active corrosion protection. It is achieved because a part of the chromate from the pretreatment solution is absorbed by the porous conversion coating formed on the metal surface. The chromate ions can be later released to the defects in the coating repassivating the metal surface in the scratch and leading to the healing of the corroded area. This feature made the chromate-based conversion coatings to be the most popular all over the world for the pretreatment of various metallic substrates. However, as it was discussed above, chromates are strongly carcinogenic, and so must be substituted by other conversion coatings. The active anticorrosion ability is also very desirable for these possible environmental-friendly candidates.

The following part shortly overviews the attempts to develop new conversion coatings with self-healing ability based on the effect of incorporated inorganic or organic corrosion inhibitors.

Molybdates, vanadates, permanganates, and tungstates were the first candidates tried as alternatives for chromates since they are oxidants and have chemical elements from groups VI and VII of the periodic table similar to chromium [54]. Molybdenum and tungsten belong to the same group and vanadium and manganese are the nearest neighbors.

The use of molybdates as corrosion inhibitors has been documented in 1976 [55]. Good results were obtained on zinc surfaces using an ammonium molybdate and ammonium chloride passivation bath [54]. Molybdate conversion coatings can be obtained by simple immersion or electrochemically under cathodic polarization.

The nature of acids and pH of the baths seem to play crucial role on the quality of the deposited layer [56]. Sulfuric, nitric, and phosphoric acids are currently used [54, 56, 57]. The formation of the conversion coatings providing corrosion inhibition is usually performed within the pH 1–6, for different acids. No conversion coating is formed at pH values above 6. Other deposition conditions such as temperature, additives, concentration, and dipping time can also influence the performance of the molybdate conversion coatings, but a systematic study has not been made to date [56]. In the case of a molybdate conversion coating on zinc, it was revealed that the best performance can be achieved when phosphoric acid is used in the bath. The self-healing effect is clearly demonstrated on the samples treated by this solution. This was in accord with an increase in impedance during the first few hours of immersion as in the case of chromate-based conversion coatings. But, the behavior of the coatings produced in H_2SO_4 and HNO_3 baths shows that the deposited conversion layers did not have well-defined self-healing due to a lack of free molybdate in the formed layer. Monticelli *et al.* tested the effects of molybdates on the corrosion of aluminum using electrochemical noise analysis (ENM) [58]. They found that molybdates act as oxidizing inhibitors, and the main inhibiting effect was due to an adsorbed layer acting as a barrier to chloride ions. The inhibition mechanism of the molybdate films produced in phosphoric acid baths was similar to that of chromate, although chromate treatments used as a reference still had a better anticorrosive performance when compared with the tested molybdate coatings [56].

Vanadates were also suggested as alternative pretreatment candidates aiming in self-healing conversion coating. Vanadate-based pretreatments were tested on different metallic substrates. Guan *et al.* developed a conversion coating for AA2024, which is formed at ambient temperatures during several minutes of immersion in aqueous vanadate-based solution [59]. Auger Electron Spectroscopy/X-Ray Photoelectron Spectroscopy (AES/XPS) methods show that conversion coatings formed by a 3-min immersion are 300–500 nm thick and comprise a mixture of vanadium oxides and other components from the coating bath. This treatment increases the pitting potential and decreases the rate of oxygen reduction. Analytical study indicates that vanadate is released into the solution upon exposure providing active corrosion protection and self-healing ability during immersion in chloride electrolyte. A study of inhibiting mechanism of vanadates confirmed that it works as a cathodic inhibitor sufficiently decreasing the oxygen reduction reaction without any influence on the anodic process [60]. Almeida *et al.* used vanadate baths (1 M, 20 °C, 300 s, and pH 5 adjusted with H_2SO_4) to prepare a conversion coating on a zinc substrate [54, 61]. The corrosion protection in this case was far below than that of chromate or molybdate baths. However, in case of magnesium alloy (AZ61) it can significantly improve the corrosion resistance [62]. According to a potential dynamic polarization test, the corrosion resistance of a vanadate-based conversion coating is higher to that of cerium- and phosphate-based conversion coatings on the same substrate.

Other oxidizing compounds that were suggested for preparation of self-healing conversion coatings are *manganates*. These pretreatments were also tested on different metallic substrates such as magnesium alloys [63], aluminum alloys [64], and zinc [54, 61]. The corrosion protection behavior of permanganate-based no-rinse conversion coatings on AA3003-H14 aluminum alloy shows that they are promising as an alternative to toxic chromate-based coatings [64]. Additionally the performance can be improved by doping the pretreatment solution with aluminum nitrate. The permanganate/phosphate-based conversion coating obtained on magnesium alloy is composed of MgO, $Mg(OH)_2$, $MgAl_2O_4$, Al_2O_3, $Al(OH)_3$, MnO_2, or Mn_2O_3 and amorphous oxyhydroxides. The DC polarization test showed that the presented conversion film for AZ alloys has equivalent corrosion protection potential as the JIS H 8651 MX-1 (similar to Dow No. 1) chromate-based method [63].

The presented overview of oxidizing pretreatments shows that the resulting conversion films improve corrosion protection to some extent and provide self-healing ability in many cases. However, all the overviewed conversion coatings based on oxidizing species unfortunately did not show a good enough anticorrosive behavior to effectively substitute the chromate conversion coatings [61].

Rare earth compounds were also suggested as possible alternatives for active conversion coatings due to their ability to form insoluble hydroxides that enables them to be used as cathodic inhibitors [65]. They have low toxicity and their ingestion or inhalation has not been considered harmful [66]. Moreover, the rare earth compounds, especially cerium, are economically competitive products because they are present in high amounts in nature [67]. The Ce-based conversion coatings can be applied by full immersion and electrochemical activation methods.

Cerium-containing treatments are very attractive for aluminum alloy, which is very susceptible to a localized corrosion attack resulting from the presence of active cathodic intermetallics [68–73]. The first attempts to obtain cerium-containing conversion coatings by full immersion were made by Hinton *et al.* [74]. In this research, coatings were formed on AA7075 using 1000 ppm $CeCl_3$ solutions during 90 h of immersion. Unfortunately, such long time periods make these immersion treatments commercially unattractive. The corrosion protection of aluminum–lithium alloys AA2090 and AA8090 after immersion treatment in lanthanide salts studied in Ref. 75 showed a decrease in pitting corrosion. However, the protective effect of this coating was limited to a very short period.

A more rapid film deposition method, named *cerating*, has been developed and patented by Hinton and Wilson [76]. In this process, the metallic alloy is immersed in an aqueous cerium salt solution containing both oxidizing agents and organic additives. The cerating method has been successfully applied not only to aluminum alloys but also to other metallic substrates, such as zinc, galvanized steels, stainless steels, cadmium, and magnesium. However, the corrosion performance of these conversion coatings in accelerated corrosion tests was not satisfactory.

Further improvements of this method resulted in a new patented treatment called *cerate coating* [77]. The cerate coating is obtained by immersing the metallic alloy for 10 min in an aqueous solution of cerium chloride with 0.3% by-volume hydrogen peroxide content. This process when applied to an AA2024 aluminum alloy, leads to a remarkable increase in pitting nucleation resistance. The role of hydrogen peroxide in cerate process consists in the formation of Ce(III) peroxo complexes such as $Ce(H_2O_2)^{3+}$ as an initial step, followed by deprotonation, oxidation, and precipitation to form peroxo-containing Ce(IV) species such as $Ce(IV)(O_2)(OH)_2$ [70].

The second group of pretreatments is based on the electrochemical activation under galvanostatic cathodic currents in solutions with cerium ions. However, an improvement in pitting corrosion resistance is lower than that of the full immersion method [66].

In case of aluminum alloys, the cerium-based conversion coatings can provide some self-healing ability. The healing of a defect originates from the unreacted cerium(III) ions that are entrapped by porous conversion coatings. Cerium cations diffuse into the metal surface when a defect is created and form insoluble deposits in cathodic places hindering the overall corrosion activity [69]. A similar self-healing mechanism was also revealed by Aramaki on zinc substrates treated with various modified cerium-based conversion coatings [78–81]. The Ce-containing conversion coatings were formed in different baths doped with phosphoric acid. In the case of undoped Ce $(NO_3)_3$ solution, a thin layer ($\sim$50 nm in thickness) of hydrated cerium oxide is formed [78]. This layer acts as a barrier for oxygen diffusion hindering cathodic activity on the metal surface. Moreover, when an artificial scratch is formed and the treated surface is immersed in a corrosive electrolyte, the Ce^{3+} species migrate to the scratch repassivating the metal surface. The same kind of conversion coating is also formed on galvanized steel covered with zinc layer [82, 83]. Phosphate modified cerium-based conversion coatings on zinc confers even more effective self-healing ability [80, 81]. The cerium-based conversion coatings were also applied to other metallic alloys, such as steel [84, 85] and magnesium alloys [86, 87], providing additional corrosion protection.

In summary, the different RE-based treatments discussed above enhance the corrosion resistance of various metallic substrates. However, data reported to date suggest that conversion coatings based on lanthanide compounds present various drawbacks that limit their application on an industrial scale.

Currently, the widely used Cr-free conversion coatings are based on the phosphatation process [53] especially for steel, which is still the main metal used for engineering needs. However, the *phosphate-based treatments* do not confer any sufficient self-healing ability to the coating system. Therefore, the idea to impart self-healing properties to the phosphate conversion coatings (PCC) seems to be very attractive. To achieve this goal, the porous structure of PCC can be doped with organic or inorganic corrosion inhibitors during or right after the preparation [88]. For example, the addition of zinc cations to a phosphating bath is quite natural because they can remarkably enhance the protective properties of many anionic inhibitors, including phosphates. Such an effect is due to the ability of zinc to form water-insoluble hydroxides, which precipitate on the steel surface decelerating the cathodic reaction. Zinc additives are especially effective during the cathodic polarization of steel in corrosive waters because of their inclusion into the salt deposits. Among inorganic inhibitors, the molibdate additives to PCC demonstrate the most effective corrosion protection. Organic inhibiting species can also be introduced to the PCC. In general, many organic compounds are known to be used as accelerators in the application of thick PCC in order to activate the metal surface. Among oxidative organic accelerators of phosphating, attention should be given to nitroarenes, including such known inhibitors of metal corrosion as nitrobenzoates, nitrophenols, nitrochromopyrazole, and so on. For example, the use of picric acid and 4-nitrosodimethylaniline confer PCC with excellent uniformity and high protection properties [88].

A porous reservoir layer on the metallic surface can also be created using a sol–gel route. A novel approach aimed at developing of a nanoporous reservoir for storing of corrosion inhibitors on the metal/coating interface has been proposed by Lamaka *et al.* [89]. A porous titanium layer on the surface of AA2024 aluminum alloy was developed using template-based synthesis controllably hydrolyzing titanium alkoxide in the presence of template agent. The reservoir comprises titanium nanoparticles, which are self-assembled, forming a cellular network that replicates the surface structure of the etched alloy, as shown in Figure 4.6. The formation of network-like oxide-based structure with a highly developed surface area on the metallic substrate opens a great opportunity for loading of this layer with active substances. For this reason, the alloy with the applied porous nanostructured film was, then, immersed to an alcoholic solution of *n*-benzotriazole, which is a known corrosion inhibitor for AA2024 [90]. After the loading with inhibitor, the substrate was coated with hybrid sol–gel coating, and then tested for corrosion protection. This novel pretreatment shows enhanced corrosion protection compared with an undoped sol–gel film.

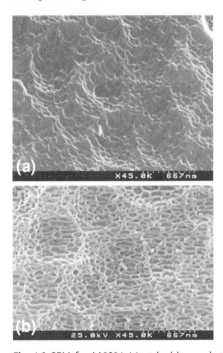

Fig. 4.6 SEM for AA2024 (a) etched bare substrate and (b) etched alloy coated with TiO_x film. Reprinted from [89] with permission from Elsevier.

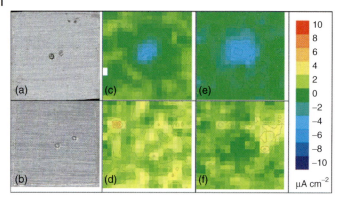

Fig. 4.7 SVET maps of the ionic currents measured 200 μm above the surface of the AA2024 coated with (c and e) sol–gel film (I) and (d and f) sol–gel film pretreated with inhibitors-loaded TiO₂ layer (IV). (a) and (b) present microphotos of samples I and IV with artificial defects. Scanned area: 2 mm × 2 mm. Reprinted from [91] with permission from Elsevier.

The pretreatment formed by the nanostructured titanium reservoir layer covered with the hybrid film demonstrates well-defined self-healing ability leading to effective long-term active corrosion protection [91]. The scanning vibrating electrode technique (SVET) was employed to study the corrosion protection and the self-healing ability of the two-layer systems on microlevel. SVET method is able to measure, along the coated surface, the local distribution of anionic and cationic fluxes which correspond to cathodic and anodic corrosion currents, respectively. Thus, the ionic fluxes in the defect can be directly correlated to the corrosion activity. The measurements of local corrosion currents evolution in the zones of artificial defects created on the coated samples are shown in Figure 4.7. Well-defined cathodic activity due to active corrosion processes appears in the defects formed on the surface of AA2024 treated only with sol–gel film, while no corrosion activity was revealed even after 24 h on the alloy pretreated with inhibitor-loaded porous TiO₂ and covered with sol–gel film. Thus, the nanostructured porous character of the titanium layer provides a high effective surface area for the adsorption of the inhibitor. The adsorbed inhibitor is leached out from the porous pretreatment layer to the defect, thereby healing it. The developed surface, formed by the self-assembled layer, also offers a good adhesion between the oxide and the sol–gel film due to the high contact area between the two phases. Moreover, the employment of this reservoir approach avoids the negative effects of the inhibitor on the stability of the sol–gel coating.

The short overview of the conversion coatings demonstrates that a number of new promising environmental-friendly pretreatments are developed to date. Some of them not only provide good adhesion between the metal surface and the polymer coating but also confer additional self-healing ability when active inhibiting species are stored in the reservoir created by porous conversion coatings. However, the conversion coatings are usually composed of thin submicron films, which have a very limited storage capacity for the inhibitor. Thus, the inhibiting species released from this layer are not able to provide a long-term healing of the defect. Therefore, the active inhibiting components should also be added to the thick primer or even to the topcoat layers ensuring a sufficiently higher loading and, consequently, a more effective self-healing effect. The following sections of this chapter are particularly devoted to the discussion on different ways of introducing the inhibitor to various organic, inorganic, and hybrid anticorrosion coatings.

4.3.3
Protective Coatings with Inhibitor-doped Matrix

The easiest way to introduce a corrosion inhibitor to the coating is just a simple mixing with the coating formulation. However, the simple mixing, in fact, is not so simple because it can raise many problems if some important factors are not taken into account. At first, the inhibiting species are effective only if their solubility in the corrosive environment is in the right range. Very low solubility leads to a lack of active agent at the substrate interface and, consequently, to a weak self-healing activity. If the solubility is too high the substrate will be protected but only for a relatively short period since the inhibitor will be rapidly leached out from the coating. Another disadvantage, which can appear due to the high solubility, is the osmotic pressure that forces water permeation leading to blistering and delamination of the protective coating. Figure 4.8 clearly shows how the solubility of the corrosion inhibitor influences the adhesion of the coating on the metal surface. A high solubility of the inhibiting species dispersed in the matrix causes very fast complete stripping of the polymer film from the substrate [10]. Another important problem can appear when the corrosion inhibitor can chemically interact with the components of the coating formulations weakening the barrier properties of the final coating. Degradation of the barrier properties resulting from the addition of an inhibitor to the coating system is the main problem hampering

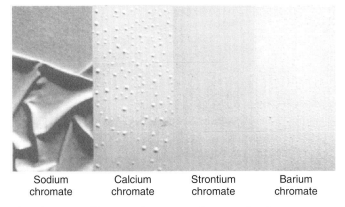

| Sodium chromate | Calcium chromate | Strontium chromate | Barium chromate |

Fig. 4.8 Osmotic blistering of organic coatings as a function of corrosion inhibitor pigments' solubility. Reprinted from [10] with permission from Elsevier.

the development of active corrosion protection systems. Moreover, interactions between the inhibitor and the components of the coating can lead to a complete deactivation of its inhibiting activity. Therefore, a direct dissolution of an inhibitor in the formulations of organic coatings is not used in practice.

The situation is different in case of hybrid organosiloxane-based films. Several attempts to produce self-healing hybrid films doped with organic and inorganic corrosion inhibitors are recently reported in the literature. Thin hybrid films are suggested, depending on thickness, as alternative pretreatments or primers for different metallic substrates. Incorporation of inorganic or organic corrosion inhibitors into the hybrid films can significantly improve the corrosion protection. The results of several works have showed that the incorporation of cerium dopants into sol–gel coatings enhances corrosion protection for aluminum alloys, magnesium alloys, galvanized steel, and stainless steel [92–102]. Voevodin *et al.* investigated corrosion protection properties of epoxy-zirconia sol–gel coatings incorporated with nonchromate inhibitors of Ce $(NO_3)_3$, $NaVO_3$, and Na_2MoO_4. The sol–gel coatings with cerium dopants perform at least as good as the undoped epoxy-zirconia films [95]. The critical concentration of the cerium inhibitor is 0.2–0.6 wt%. A higher concentration could lead to the formation of defects in the network of the sol–gel film [94]. However, the sol–gel films with $NaVO_3$ and Na_2MoO_4 did not provide adequate corrosion protection due to a decrease in the sol–gel network stability. The positive effect was achieved only when molibdate anions were added in a "bound" form, as amine molybdate, in contrast to a form of "free" ions. The bounding with the amine protects molibdate ions from interaction with the components of the hybrid matrix [103].

Organic inhibitors can also be incorporated into the sol–gel matrix in order to improve corrosion protection of metallic substrates [104]. An additional inhibition effect was revealed when phenylphosphonic acid was introduced in a hybrid sol–gel film containing phenyl groups [105].

In several cases, the release of organic molecular species from the hybrid sol–gel matrix can be described by the pH-dependent triggered release mechanism [106]. The triggering of the desorption processes can provide an "intelligent" release of the corrosion inhibitor only in places of local pH changes originating from localized corrosion processes. However, ionizable inhibitors show far weaker effect than that of nonionizable inhibitors since the former are too strongly attached to the sol–gel matrix and, thus, cannot be released during corrosion [25]. Khramov *et al.* studied the corrosion protection properties of hybrid films incorporated with mercaptobenzothiazole (MBT) and mercaptobenzimidazole (MBI) as corrosion inhibitors [107]. Figure 4.9 shows the current density distribution maps obtained by SVET for Al/Cu artificial defects on AA2024 coated with undoped and doped, with MBI, sol–gel films. After 3 h of immersion, the current peak in the case of the hybrid film with inhibitor is almost 25 times lower than that for the undoped one, showing the inhibition of the corrosion processes by the MBI molecules. Van Ooij *et al.* doped organosilane films with tolyltriazole and benzotriazole inhibitors. The organic inhibitor, tolyltriazole, added to the silane film, improved the overall corrosion resistance of the AA2024-T3 alloy, but did not impart a self-healing

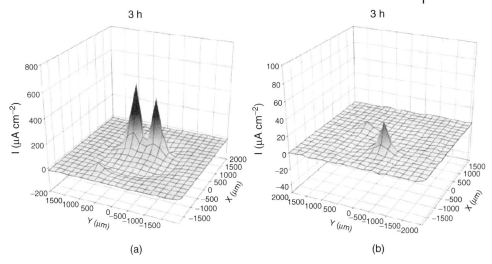

Fig. 4.9 Current density distribution maps for AA2024 coated with (a) hybrid film and (b) hybrid film doped with mercaptobenzimidazole. Reprinted from [107] with permission from Elsevier.

effect [92]. An organic corrosion inhibitor, tetrachloro-p-benzoquinone (chloranil), has also been incorporated into a hybrid organosiloxane/zirconia sol–gel matrix in order to improve corrosion protection [108]. The addition of a high content of chloranil leads to a disorganization of the sol–gel matrix and, consequently, to a low corrosion protection. However, the incorporation of lower concentrations of chloranil has implicated homogeneous structures increasing the protective properties of the sol–gel coatings.

A new approach for active corrosion protection of aluminum alloys was recently suggested by Jakab et al. [109, 110]. In this case, a metallic alloy film is used as a self-healing anticorrosive coating with multiple functionalities. A new amorphous Al–Co–Ce alloy coating was synthesized and applied on the aluminum surface. At first, such metallic coatings can function by providing sacrificial anode-based cathodic protection. But, the fact that the coating contains cerium and cobalt enables it to generate corrosion inhibiting species. The Co^{2+} and Ce^{3+} cations are released during anodic dissolution of the metallic coating. Dissolved cerium and cobalt cations can diffuse, then, to the metallic substrate surface and form insoluble hydroxide precipitates covering the cathodic zones on the aluminum alloy. This work shows one of the first examples of a pH-controlled release of corrosion inhibiting ions from an amorphous metallic coating where the pH change, which triggers the release, is a consequence of the onset of corrosion. This corrosion inhibition strategy provides further corrosion protection beyond the traditional roles of barrier and sacrificial cathodic protection using a metal coating [109]. The level of protection conferred by these alloys was found to be higher compared to the conventional Alclad coatings in terms of both polarization and maximum scratch size protected. The metallic amorphous coating can provide self-healing of scratches up to 2500 μm of exposed AA2024-T3 via chemical inhibition mechanism [111].

Interesting example of a coating that can release inhibiting species on demand using redox process as a trigger is described in [112]. Galvanic reduction of 2,5-dimercapto-1,3,4-thiadiazole polymer in a conducting carbon paste releases its monomer anion. The monomer anion, in turn, exhibits a very good inhibition efficiency for cathodic oxygen reduction process. This chemistry forms the basis for a "smart" self-healing material that releases an inhibitor when the material is coupled to a base metal as in the case of a coating with a defect. No release would occur in the absence of the defect due to a lack of the reducing force of the base metal.

Another inorganic $Al_2O_3 \cdot$ Nb nanocomposite self-healing coating for corrosion protection of iron-based substrates was recently developed by Yasuda *et al.* [113]. The idea to use very thin and dense Al_2O_3 layers on different metallic substrates as an artificial passive film is already known since the beginning of 1990s'. However, the films comprising only the oxide have no self-healing ability causing fast development of localized corrosion at the damaged site. Therefore, it is important for such films to possess a self-healing ability. Self-healing properties were achieved by introducing a metallic component into the oxide films. The oxidation of the metallic component can heal cracks in the composite films. The addition of metallic niobium to the oxide increases the self-healing ability but at the same time causes undesirable increase of pinhole density. Therefore, composition-gradient films with the content of Nb increasing from the film surface to the substrate interface were developed. For example, a composition-gradient $Al_2O_3 \cdot$ Nb composite film of 0 (top)−96 (bottom) Nb shows low pinhole defect density and high self-healing ability.

In summary, a direct addition of a corrosion inhibitor to the coating formulation in some cases confers additional active corrosion protection and self-healing ability, especially when a metallic film is used as a sacrificial reservoir for corrosion inhibitors. But, very often the inhibitor dissolved in the polymer coating causes weakening of the barrier properties and, consequently, of the overall corrosion protection. Therefore, other strategies of inhibitor introduction should be used in these cases to isolate the inhibitor from the coating components.

4.3.4
Self-healing Anticorrosion Coatings based on Nano-/Microcontainers of Corrosion Inhibitors

New active multifunctional coatings should rapidly release the active inhibiting species on demand within a short time after the changes in the environment or in the coating integrity. Recent developments in surface science and technology show new concepts for fabrication of self-healing coatings through integration of nanoscale containers (carriers) loaded with active inhibiting compounds into existing conventional coatings. This design leads to completely new coating systems based on "passive" host–"active" guest structures. As a result, nanocontainers are uniformly distributed in the passive matrix keeping the active species in a "trapped state". This removes the possibility of excessive inhibitor leaching or pigment-induced osmotic blistering, the problems which may affect the inhibitor salts if they are too soluble or prepared with too small particle size. When the local environment changes or if a corrosion process starts in the coating defect, the nanocontainers respond to this signal and release the immobilized active material [114, 115]. The concepts of the inhibitor nanocontainers can be classified into two groups, namely: encapsulation with different types of shells and immobilization on the surface or inside some carriers. In this section, the main approaches of inhibitor encapsulation and immobilization on different carriers are overviewed in terms of their applicability for the self-healing anticorrosion coatings.

4.3.4.1 Coatings with Micro-/Nanocarriers of Corrosion Inhibitors

A quite simple approach of an inhibitor entrapment is based on *complexation* of organic molecules by cyclodextrin [25, 116]. At the beginning, this concept was suggested for controllable drug delivery systems [117]. Cyclodextrins complexation agents that can play the role of hosts forming inclusion complexes with various organic guest molecules that fit into the size of the cyclodextrin cavity. Organic aromatic and heterocyclic compounds are usually the main candidates for the inclusion complexation reaction [118]. Several organic heterocyclic compounds act as inhibitors for various metallic substrates. So, cyclodextrins can be effectively used for immobilization of these species. Two organic corrosion inhibitors, MBT and MBI, were selected by Khramov *et al.* to add to a hybrid sol–gel film used for corrosion protection of AA2024 aluminum alloy. MBT and MBI were introduced to the sol–gel formulations as inclusion host–guest complexes with β-cyclodextrin. The hybrid films doped with corrosion inhibitors provide superior corrosion protection when compared to the undoped ones. Moreover, the coatings doped with inhibitors in entrapped form outperform those made by a simple addition. Figure 4.10 shows the appearance of the artificial scratch and the impedance spectra on the sol–gel coated AA2024 after four weeks of immersion in dilute Harrison's solution. The sample doped with cyclodextrin–inhibitor complex keeps the scratch almost shiny without visible corrosion products on it in contrast to the hybrid film directly doped with MBI where white corrosion products cover the defect. The higher impedance spectra after such a long corrosion test also confirms the superior corrosion protection in case of the film doped with immobilized MBI. Thus, the formulations that contain β-cyclodextrin demonstrate superior corrosion protection properties as the complexation equilibrium results in the slowed release of the inhibitor and its continuing delivery to the corrosion sites followed by self-healing of corrosion defects. However, the complexation with cyclodextrin confers only a prolonged release of the inhibitor without a delivery of it on demand as a response to any external stimuli.

Another entrapment concept is based on the use of *oxide nanoparticles*, which can play a role of nanocarriers of corrosion inhibitors absorbed on their surface. The oxide nanoparticles by themselves are reinforcements for the coatings formulations, as their addition leads to enhanced barrier properties [119–122]. Incorporation of

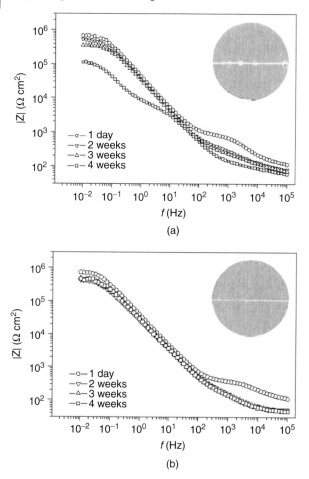

Fig. 4.10 The electrochemical impedance spectra for scribed hybrid coatings at different immersion times in dilute Harrison's solution without inhibitor (a) and with MBI/h-cyclodextrin complex (b). In the inset: optical images of the samples after four weeks of immersion. Reprinted from [25] with permission from Elsevier.

the nanoparticles into hybrid sol–gel formulations leads to an additional improvement of the barrier properties, for example, due to an enhanced thickness and a low crack ability of such composites [122, 123]. Moreover, an additional active corrosion protection and a self-healing ability can be achieved when the oxide nanoparticles are doped with a corrosion inhibitor. Immobilization of an inhibitor, Ce^{3+} ions, on the surface of the zirconia nanoparticles can be achieved during the synthesis of the nanosol controllably hydrolyzing the precursor by a Ce-containing aqueous solution [124, 125]. The resulting sol mixed with hybrid sol–gel formulation leads to nanocomposite hybrid coatings containing oxide nanocontainers of cerium ions. The high surface area of the carriers, which results from small diameter of the

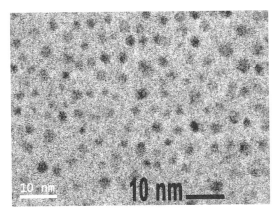

Fig. 4.11 Transmission electron microscopy (TEM) image of hybrid sol–gel film containing zirconia-based nanoreservoirs.

nanoparticles ($\sim$4 nm as shown in Figure 4.11), provides sufficient loading capacity for the inhibitor. The prolonged release of inhibitor from the surface of oxide nanoparticles provides an enhanced long-term corrosion protection for AA2024 aluminum alloy compared to the case when the inhibitor is directly added to the sol–gel matrix. Moreover, the use of oxide nanocarriers prevents the negative effect of cerium ions on hydrolytical stability of hybrid sol–gel coating [125]. The inhibiting ions can also be immobilized on the surface of commercially available nanoparticles by simple adsorption in an inhibitor containing solution. Silica- and ceria-based nanocarriers obtained by this method provide additional active corrosion protection to an organosilane coating applied to galvanized steel [126].

Nanoparticle surface as inhibitor carrier also adsorb the inorganic corrosion inhibiting anions onto the particle surface by an ion-exchange mechanism [127]. The corrosion inhibitors are released from the particle surfaces by a subsequent ion-exchange with anions (e.g. chlorides, sulfates, and sodium ions) transported into the coating from the environment via water penetrating through the coating. However, this mechanism can also lead to an undesirable release of the inhibitor by the presence of harmless ions in the surrounding environment, during, for example, cleaning, and not only by the ions resulting from the corrosion process.

Organic corrosion inhibitors can also be immobilized on the surface of nanoparticles. Chemical anchoring of an organic inhibitor to aluminum oxyhydroxide nanoparticles through carboxylic bonds was employed in protective coatings for aluminum, copper, nickel, brass, and bronze substrates [128]. Hydroxide ions generated from the corrosion of these metals trigger the release of the corrosion inhibitors from the particles at pH 9. Thus the release of the inhibitor starts only by the corrosion process, which prevents undesirable leakage of the inhibitor from an intact coating during exploitation. High specific surface area of oxyhydroxide nanocarrier (at least $100\,\mathrm{m}^2\,\mathrm{g}^{-1}$) allows a higher quantity of corrosion inhibitor to be delivered into the damaged part of the coating.

Even higher loading capacity can be achieved when porous fillers with *hollow cellular structure* are loaded with organic and/or inorganic inhibitors [129]. The hollow cellular structure material may be represented by diatomaceous earth, zeolite, or carbon. *Zeolite* particles are also attractive carriers because the cations in their structure are rather loosely held and can readily be exchanged for the inhibiting cations in the contact solution [130, 131].

Inhibiting inorganic cations can also be incorporated as exchangeable ions associated with *cation-exchange solids* [132–134]. The advantage of this approach is that the cation-exchange pigment is completely insoluble avoiding osmotic blistering [134]. Calcium(II) and cerium(III) cation-exchanged bentonite anticorrosion pigments are prepared by exhaustive exchange of naturally occurring Wyoming bentonite. The cation exchange was carried out by a repeated washing with aqueous solutions of cerium(III) chloride and calcium(II) chloride to produce bentonites containing 31 500 ppm exchangeable cerium(III) and 13 500 ppm exchangeable calcium(II), respectively. Bentonite clays are a form of montemorillonite and exhibit intrinsic cation-exchange properties. They are constituted by stacked negatively charged aluminosilicate layers. The negative charge of these layers is compensated by the cations intercalated between the aluminosilicate sheets. The interaction between the sheets and the exchangeable cations is purely electrostatic. Therefore, the principal exchangeable cations (sodium(I) and calcium(II)) can be easily exchanged in the laboratory either by passing a suspension of bentonite through a cation-exchange column or by repeatedly washing the bentonite with a solution containing the desired cation. The calcium(II) and cerium(III) bentonite pigments have been dispersed (19% pigment-volume-content (PVC)) in polyester primer layers applied to pretreated hot dip galvanized steel [134]. Salt spray test studies have shown that the calcium(II) and cerium(III) bentonite pigments are effective inhibitors of corrosion driven coating delamination at the cut edge of the polymer-coated steel. The coatings doped with a Ce^{4+}-modified ion-exchanged pigment also demonstrate promising corrosion inhibition by cerium cations transported to the active defect sites on bare aluminum surfaces.

The mechanism of corrosion inhibition using the Ce-doped cation-exchange pigment is shown in Figure 4.12. The pigment remains entrapped inhibiting cations between the aluminosilicate sheets. The cations from the corrosive medium interact with the pigment particles when the corrosive electrolyte penetrates into the defect in the coating while exchanging the inhibiting cations. The Ce^{3+} ions diffuse, then, to the metallic surface and react with hydroxyl ions originating from cathodic oxygen reduction. Consequently, the insoluble cerium hydroxide precipitates block the cathodic sites on the metal surface reducing the corrosion activity and healing the defect. However, as in the case of oxide nanoparticles and zeolites, the release of inhibitor here is based on the prolonged leaching and is not triggered by some specific corrosion-related stimuli.

A different situation appears when *anion-exchange pigments* are used to immobilize anionic inhibitors [135–138]. The release of inhibitor anions can be provoked in this case by aggressive corrosive chloride ions. The anion-exchange pigment can play a double role, absorbing the harmful chlorides and releasing the inhibiting ions

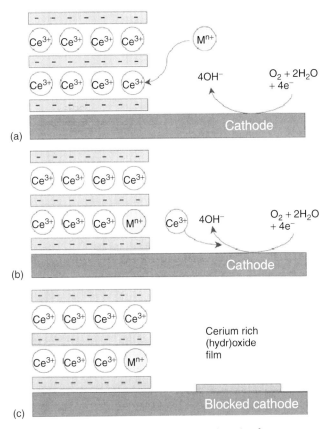

Fig. 4.12 Schematic illustration of the potential mode of operation of a cation-exchanged bentonite inhibitor pigment on a corroding galvanized surface. Reprinted from [134] with permission from Wiley-VCH Verlag GmbH & Co. KGaA.

in response. Even only the "trap" function can provide the additional anticorrosion effect in filiform corrosion tests when noninhibited pigments with carbonates and sulfates are used in organic coatings [138]. Absorption of chlorides from an aggressive electrolyte in the vicinity of a defect decreases the aggressiveness of the corrosive medium decreasing the rate of the corrosion processes. The use of an inhibiting anion-exchanged pigment additionally can confirm the active feedback conferring a self-healing ability.

Layered double hydroxide compound, hydrotalcite (HT), is a host–guest structure and can be used as an effective anion exchanger. The structure consists of a host, positively charged Al–Zn hydroxide layers, separated by layers of anions and water (Figure 4.13). The positive charge originates from the substitution of Al^{3+} on Zn^{2+} sites in the structure. On contact with an aggressive environment containing chlorides, an exchange reaction will occur. In this reaction, the inhibitor anions are released and the chlorides are adsorbed into the HT gallery. The exchange reaction is chemical in nature and is governed by the equilibrium constant: $HTInh + NaCl$

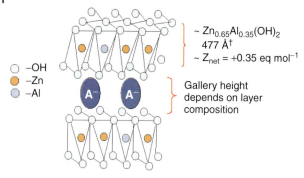

Fig. 4.13 Schematic illustration of the layered structure in hydrotalcite compounds. The structure consists of alternating layers of positively charged mixed metal hydroxide sheets and anions. Reprinted from [137] with permission from Elsevier.

(aq) $\leftrightarrow$ $HTCl$ + NaInh (aq) [137]. Buchheit *et al.* synthesized Al–Zn–decavanadate HT pigments and added them to the epoxy-based coatings applied to AA2024 aluminum alloy. Results from corrosion tests illustrate an additional corrosion protection by the HT pigment due to the decavanadate release accompanied by the uptake of the chloride ion in the exchange reaction [136].

The ability of HT to act as a generic anion delivery system creates a possibility to immobilize organic anionic inhibitors intercalating them between hydroxide layers [139]. An HT pigment doped with benzotriazolate, ethyl xanthate, and oxalate anions was developed and tested in terms of inhibiting efficiency with respect to filiform corrosion on organic coated AA2024. Inhibitor efficiency increases in the order ethyl xanthate < oxalate < benzotriazole. However, it is not as efficient as chromate-based pigment and its practical usefulness remains unproven. Nevertheless, HT pigments doped with organic inhibitor anions appear worthy of further investigation as inhibitors for self-healing anticorrosion coatings.

4.3.4.2 Coatings with Micro-/Nanocontainers of Corrosion Inhibitors

Various approaches of corrosion inhibitor immobilization on different nano-/microparticulate carriers were discussed in the previous subsection. However, only the described approach cannot be used to provide prolonged release of inhibitor and protect it from the interaction with the coating formulation. Another strategy is based on the use of different encapsulation techniques when a protective shell on a core containing inhibitor is created.

The *encapsulation* of active healing agents for protective coatings was already discussed in Section 4.2.2. A corrosion inhibitor can be encapsulated together with a polymerizable healing agent [27, 28].

Yang et al. encapsulated triazole inhibitor using *plasma polymerization* (PP) to produce PP-perfluorohexane and PP-pyroll layers employing radio frequency (RF) plasma discharge [140]. The plasma-treated triazole was used as a pigment in a water-based epoxy coating slowly releasing the inhibitor and providing a long-term corrosion protection. In both cases, the release of the inhibitor from the capsule is possible only when it is mechanically damaged. The damaged capsule releases the entire active agent very fast in a noncontrollable way.

A very interesting alternative that allows a controllable leaching triggered by a corrosion-related stimulus is the use of *layer-by-layer (LbL) assembled shells*. Nanocontainers with regulated storage/release of the inhibitor can be constructed with nanometer-scale precision employing the LbL deposition approach [141]. With such a step-by-step deposition of oppositely charged substances (e.g. polyelectrolytes, nanoparticles, and biomaterials) from their aqueous and nonaqueous solutions on the surface of a template material, the LbL shells were assembled and investigated as perspective materials for different applications [142, 143]. LbL films containing one or several polyelectrolyte monolayers assembled on a surface of a sacrificial template possess the controlling of the shell permeability toward ions and small organic molecules. The storage of corrosion inhibitors in the polyelectrolyte multilayers has two advantages: isolation of the inhibitor avoiding its negative effect on the integrity of the coating and providing an intelligent release of the corrosion inhibitor regulating the permeability of polyelectrolyte assemblies by changing local pH and humidity. The change in pH is the most preferable trigger to initiate the release of the inhibitor since, as well known, corrosion activity leads to local changes in pH near cathodic and anodic defects. Thus, a "smart" coating containing polyelectrolyte containers can detect the beginning of the corrosion and start the self-healing process in the corrosion defect [114, 115].

The possibility to create such a smart self-healing anticorrosion coating based on LbL assembled nanocontainers was recently demonstrated by Zheludkevich *et al.* [144, 145]. Silica nanoparticles were used as a template and benzotriazole as an organic corrosion inhibitor. The LbL deposition procedure was employed involving both large polyelectrolyte molecules and small benzotriazole ones. The initial SiO_2 nanoparticles are negatively charged, so the deposition of the positive poly(ethylene imine) (PEI) was performed, in the first stage. Then, the deposition of the negative poly(styrene sulfonate) (PSS) layer was carried out. Deposition of the third inhibitor layer was accomplished in acidic media (pH 3) from a $10\,mg\,ml^{-1}$ solution of benzotriazole. The latter two deposition steps (PSS and benzotriazole) were repeated again to ensure a high inhibitor loading in the final LbL structure. The optimal number of PSS/benzotriazole bilayers that should be deposited onto silica nanoparticles is two [145]. One bilayer is not sufficient for the self-healing effect of the final protective coating while three or more bilayers drastically increase aggregation of the nanocontainers during assembly and coating deposition. The benzotriazole content in nanocontainers is equal to $95\,mg\,1\,g^{-1}$ of the initial SiO_2 particles. The assembly process of such a nanocontainer of corrosion inhibitor is schematically depicted in Figure 4.14a [146]. Nanocontainers of this type cannot prevent spontaneous leakage; however, after 60 days of aging, the nanoparticulated reservoirs still contain benzotriazole in a quantity up to 80% of the initial inhibitor loading [146].

The sol of nanocontainers containing benzotriazole was mixed with a hybrid sol–gel formulation and then applied on the surface of AA2024 aluminum alloy [144, 145]. The samples coated with LbL nanocontainers-doped hybrid film demonstrate sufficiently enhanced performance in corrosion tests in comparison

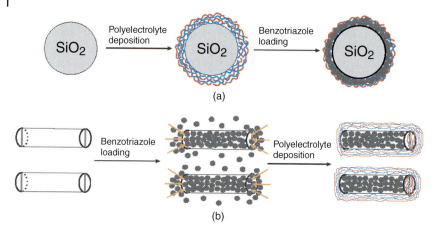

Fig. 4.14 Schematic illustration of the procedure for benzotriazole loading for (a) SiO$_2$ nanoparticulate containers and (b) halloysite nanotubes. Reprinted from [146] with permission from Wiley-VCH Verlag GmbH & Co. KGaA.

to an undoped sol–gel film or a film doped directly with free nonimmobilized benzotriazole. After two weeks of immersion in a chloride solution, the nanoreservoir-containing film is still intact, while the sample coated with hybrid coating directly doped with benzotriazole shows extensive corrosion attack with many pittings on the alloy surface (Figure 4.15) [144].

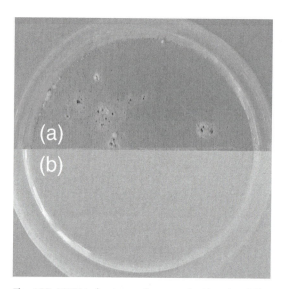

Fig. 4.15 AA2024 aluminum alloy coated with sol–gel film directly doped with benzotriazole (a) sol–gel film doped with LbL nanocontainers (b) after 14 days of immersion in 0.005 M NaCl (a) and 0.5 M NaCl (b). Reprinted from [144] with permission from American Chemical Society.

SVET experiments were performed to check if this superior corrosion protection performance of a nanoreservoir-containing coating is related to its self-healing ability. A typical current map along an intact sol–gel film is showed in Figure 4.16a illustrating the absence of local corrosion processes for both the coatings. Artificial defects (200 μm in diameter) were formed on the surface of both the coatings after 24 h of immersion in 0.05 M NaCl, as shown in Figure 4.16b and f. Well-defined cathodic activity appears in the place of the induced defect on the alloy coated with the undoped hybrid film (Figure 4.16c). This activity becomes even more intense with the immersion time (Figure 4.16d and e). A sufficiently different behavior was revealed after the defect formation on the substrate coated with the nanocontainers-doped hybrid film. No corrosion activity appears in this

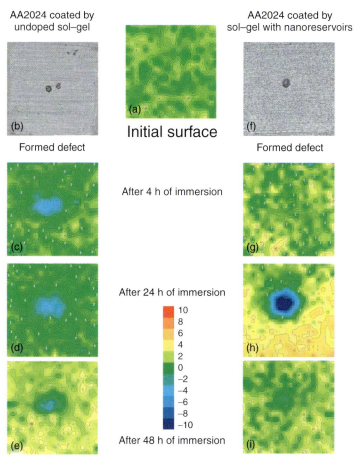

Fig. 4.16 SVET maps of the ionic currents measured above the surface of the AA2024 coated with undoped sol–gel pretreatment (a and c–e) and with pretreatments impregnated by LbL nanocontainers (g–i). The maps were obtained before defect formation (a) and for 4 h (c and g), 24 h (d and h), and 48 h (e and i) after defect formation. Scale units: I, A cm^{-2}. Scanned area: 2 mm × 2 mm. Reprinted from [144] with permission from American Chemical Society.

case after 4 h following the defect formation (Figure 4.16g). Only after about 24 h, the well-defined cathodic activity appears in the zone of the induced defect (Figure 4.16h). The rest of the surface generates a cationic flow. However, the defect becomes passivated again only 2 h later and remains healed even after 48 h (Figure 4.16i).

One can see that local corrosion activity triggers the release of a portion of benzotriazole from the nanocontainers hindering the corrosion process in the defective area. Such a "smart" self-healing effect can originate from the active interaction between the coating and the localized corrosion processes. The most probable mechanism is based on the local change in pH in the damaged area due to the corrosion processes. When the corrosion processes start, the pH value changes in the neighboring area, which opens the polyelectrolyte shells of the nanocontainers in a local area with the following release of benzotriazole. Then, the released inhibitor suppresses the corrosion activity and the pH value is recovered closing the polyelectrolyte shell of nanocontainers and terminating further release of the inhibitor, as schematically shown in Figure 4.17.

The promising results obtained on the LbL nanocontainers described above show principal possibility to use this approach for intelligent self-healing coatings. However, the nanocontainers with silica core do not provide high inhibitor loading capacity. Porous cores are more promising in this case. One of the prospective future containers can be industrially mined *halloysite nanotubes*. Halloysite was found to be a viable and inexpensive nanoscale container ($4 per kg with the supply of 50 000 t per year) for the encapsulation of biologically active molecules. The lumen of the halloysite can be used as an enzymatic nanoreactor [147]. A strong surface charge on the halloysite tubules has been utilized for designing nano-organized

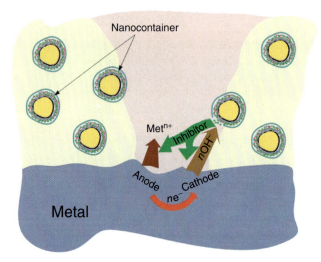

Fig. 4.17 Scheme of the controllable release of the inhibitor from the LbL nanocontainers and the "smart self-healing" process. Reprinted from [144] with permission from American Chemical Society.

multilayers using the LbL method [148, 149]. Halloysite is defined as a two-layered aluminosilicate, chemically similar to kaolin, which has predominantly hollow tubular structure in the submicron range. The neighboring alumina and silica layers create a packing disorder, which causes them to curve. As for the most natural materials, the size of halloysite particles vary within 1–15 μm of length and 10–150 nm of inner diameter depending on the deposits. The ζ-potential behavior of halloysite particles is negative at pH 6–7 and similar to the surface potential of SiO_2 with a small contribution from the positive Al_2O_3 inner surface (chemical properties on the halloysite nanotubes outermost surface are similar to the properties of SiO_2 while the properties of the inner cylinder core could be associated with Al_2O_3). However, at pH 8.5 the tubule lumen has a positive surface promoting the loading of negative macromolecules and preventing their adsorption on the negatively charged outer surface. Halloysite nanotubes are capable of entrapping a range of active agents (within the inner lumen, as well as within void spaces in the multilayered aluminosilicate shell) followed by their retention and release. Both hydrophobic and hydrophilic agents can be embedded after an appropriate pretreatment of the halloysite [150].

Cheap halloysite nanotubes as perspective nanocontainers for anticorrosion coatings with active corrosion protection were recently demonstrated in Refs [146, 151]. Halloysite nanotubes were loaded with a corrosion inhibitor 2-mercaptobenzothiazole, which is partly soluble in water and well soluble in ethanol or acetone, and then incorporated into a hybrid sol–gel coating. To prevent undesirable leakage of the loaded inhibitor from the halloysite interior, the outer surface of the 2-mercaptobenzothiazole-loaded halloysite nanotubes was modified by a deposition of several alternating polyelectrolyte multilayers (poly(allylamine hydrochloride) and PSS) as shown in Figure 4.14b. Another functionality of the polyelectrolyte shell is to provide the release of the encapsulated inhibitor in a way controlled by pH changes in the environment surrounding the halloysite nanotube [146], which will prevent a spontaneous leakage of the inhibitor and perform its release triggered by the pH changes directly in the corrosion pit. Halloysite nanocontainers showed very good upkeep characteristics—almost complete suppression of the inhibitor release with more than 90% of the initial loading retained inside the inner cavity. This can be explained by the geometrical restriction of the nanotubular container that is able to release the encapsulated material only through the polyelectrolyte-blocked edges with diameters of 20–50 nm. Aluminum alloy AA2024-T3 was taken as a model metal substrate. The results of long-term corrosion tests demonstrated superior corrosion protection performance of halloysite-doped hybrid sol–gel films compared to that of undoped coatings [151].

4.4
Conclusive Remarks and Outlook

An overview of different approaches for self-healing anticorrosion coatings is presented here. Two completely different concepts of self-healing are suggested for protective coating systems. The first one is closer to the classical understanding of self-healing and is based on the mechanisms that allow the recovery of mechanical integrity of damaged coatings. However, the different approaches, which are based on the active suppression of corrosion processes in defected areas, are also considered here as another self-healing concept.

Looking to the future of self-healing anticorrosion coatings, an idea of multilevel self-repair response seems to be the most promising. A multilevel self-healing concept would combine in the same coating system the different damage reparation mechanisms that gradually act in response to different environmental impacts. The different active components of the protective system must be able to respond to the different type and levels of impact imposed to the coating. One example of this approach is demonstrated in the work of Kumar *et al.* [27] when a polymerizable healing agent and a corrosion inhibitor were encapsulated together providing the healing of the defect by polymerization and corrosion inhibition at the same time. The synergistic protective effect originating from the combination of the different self-healing mechanisms can be achieved by incorporating different types of nanocontainers in the same coating system. These nanocontainers can be added to the same polymer film or added to different layers such as primer, clear coat, or topcoat depending on their functionality and the target of the active compounds. This approach allows the creation of protective coatings that will adequately respond to the environmental impacts providing effective self-healing and long-term service life of the anticorrosion coating systems. One can expect that the performance of the coating that is constituted by a conversion coating with active anticorrosion component, a primer doped with "smart" nanocontainers of corrosion inhibitor and a topcoat with capsules of a polymerizable healing agent, will provide outstanding corrosion protection properties and a long-term performance without the necessity of repairs.

And finally, one of the most important issues, which limit a wide use of self-healing coatings for different commercial applications, is the high cost of the suggested technologies. Only a more extensive development of these approaches and high investments to the area can lead to lower prices opening the exciting possibilities to see the "smart" self-healing coatings in our day-to-day life.

References

1 Feng, W., Patel, S.H., Young, M.-Y., Zunino, J.L. and Xanthos, M. (2007) *Advances in Polymer Technology*, **26**, 1–13.

2 Li, W. and Calle, L.M. (2006) *Smart Coating for Corrosion Sensing and Protection*. Proceedings of the US Army Corrosion Summit 2006, Clearwater Beach, Feb. 14–16.

3 Bierwagen, G.P. and Tallman, D.E. (2001) *Progress in Organic Coatings*, **41**, 201–16.

4 Chattopadhyay, A.K. and Zenter, M.R. (1990) *Aerospace Aircraft Coatings*, Federation of Societies for Paint Technology, Philadelphia.

5 Kendig, M.W. and Buchheit, R.G. (2003) *Corrosion*, **59**, 379–400.

6 Osborne, J.H. (2001) *Progress in Organic Coatings*, **41**, 280–86.

7 Twite, R.L. and Bierwagen, G.P. (1998) *Progress in Organic Coatings*, **33**, 91–100.

8 Scholes, F.H., Furman, S.A., Hughes, A.E., Nikpour, T., Wright, N., Curtis, P.R., Macrae, C.M., Intem, S. and Hill, A.J. (**2006**) *Progress in Organic Coatings*, **56**, 23–32.

9 Furman, S.A., Scholes, F.H., Hughes, A.E., Jamieson, D.N., Macrae, C.M. and Glenn, A.M. (**2006**) *Corrosion Science*, **48**, 1827–47.

10 Sinko, J. (**2001**) *Progress in Organic Coatings*, **42**, 267–82.

11 Sehgal, A., Frankel, G.S., Zoofan, B. and Rokhlin, S. (**2000**) *Journal of the Electrochemical Society*, **147**, 140–48.

12 Pride, S.T., Scully, J.R. and Hudson, J.L. (**1994**) *Journal of the Electrochemical Society*, **141**, 3028–40.

13 Trueman, A.R. (**2005**) *Corrosion Science*, **47**, 2240–56.

14 Howard, R.L., Zin, I.M., Scantlebury, J.D. and Lyon, S.B. (**1999**) *Progress in Organic Coatings*, **37**, 83–90.

15 Trask, R.S., Williams, H.R. and Bond, I.P. (**2007**) *Bioinspiration and Biomimetics*, **2**, 1–9.

16 White, S.R., Sottos, N.R., Geubelle, P.H., Moore, J.S., Kessler, M.R., Sriram, S.R., Brown, E.N. and Viswanathan, S. (**2001**) *Nature*, **409**, 794–97.

17 Cho, B.S.H., Andersson, H.M., White, S.R., Sottos, N.R. and Braun, P.V. (**2006**) *Advanced Materials*, **18**, 997–1000.

18 Allsop, N.A., Bowditch, M.R., Glass, N.F.C., Harris, A.E. and O'Gara, P.M. (**1998**) *Thermochimica Acta*, **315**, 67–75.

19 Yin, T., Rong, M.Z., Zhang, M.Q. and Yang, G.C. (**2007**) *Composites Science and Technology*, **67**, 201–12.

20 Brown, E.N., White, S.R. and Sottos, N.R. (**2005**) *Composites Science and Technology*, **65**, 2474–80.

21 Williams, G., Trask, R. and Bond, I. (**2007**) *Composites*, **38**, 1525–32.

22 Trask, R.S., Williams, G.J. and Bond, I.P. (**2007**) *Journal of the Royal Society Interface*, **4**, 363–71.

23 Pang, J.W.C. and Bond, I.P. (**2005**) *Composites Science and Technology*, **65**, 1791–99.

24 Toohey, K.S., Sottos, N.R., Lewis, J.A., Moore, J.S. and White, S.R. (**2007**) *Nature Materials*, **6**, 581–85.

25 Khramova, A.N., Voevodinb, N.N., Balbysheva, V.N. and Mantzc, R.A. (**2005**) *Thin Solid Films*, **483**, 191–96.

26 Enos, D.G., Kehr, J.A. and Guilbert, C.R. (**1999**) *A High-Performance, Damage-Tolerant, Fusion-Bonded Epoxy Coating*. Pipeline Protection Conference n° 13 (1999), Edinburgh, Scotland: 29 September - 1 October.

27 Kumar, A., Stephenson, L.D. and Murray, J.N. (**2006**) *Progress in Organic Coatings*, **55**, 244–53.

28 Sauvant-Moynot, V., Gonzalez, S. and Kittel, J. (**2007**) *Self-healing Coatings: an Alternative Route for Anticorrosion Protection*. AETOC-2007, Baiona, Spain, 18–21 April.

29 (**2005**) Nissan Develops World's First Clear Paint that Repairs Scratches on Car Surfaces. *JCNN News Summaries*. 2005, Dec. 5.

30 Sugama, T. and Gawlik, K. (**2003**) *Materials Letters*, **57**, 4282–90.

31 Hikasa, A., Sekino, T., Hayashi, Y. Rajagopalan, R. and Niihara, K. (**2004**) *Materials Research Innovations*, **8**, 84–88.

32 Tallman, D.E., Spinks, G., Dominis, A. and Wallace, G.G. (**2002**) *Journal of Solid State Electrochemistry*, **6**, 73–84.

33 Mengoli, G., Munari, M., Bianco, P. and Musiani, M. (**1981**) *Journal of Applied Polymer Science*, **26**, 4247–57.

34 Beck, F., Michaelis, R., Schloten, F. and Zinger, B. (**1994**) *Electrochimica Acta*, **39**, 229–34.

35 Kinlen, P.J., Silverman, D.C. and Jeffreys, C.R. (**1997**) *Synthetic Metals*, **85**, 1327–32.

36 Lu, W., Elsenbaumer, R.L. and Wessling, B. (**1995**) *Synthetic Metals*, **71**, 2163–66.

37 Schauer, T., Joos, A., Dulog, L. and Eisenbach, C.D. (**1998**) *Progress in Organic Coatings*, **33**, 20–27.

38 Deng, Z., Smyrl, W.H. and White, H.S. (**1989**) *Journal of the Electrochemical Society*, **136**, 2152–58.

39 Brusic, V., Angelopoulos, M. and Graham, T. (**1997**) *Journal of the Electrochemical Society*, **144**, 436–42.

40 Lacroix, J.C., Camalet, J.L., Aeiyach, S., Chane-Ching, K.I., Petitjean, J., Chauveau, E. and Lacaze, P.C. (**2000**) *Journal of Electroanalytical Chemistry*, **481**, 76–81.

41 Cocchetto, L., Ambat, R., Davenport, A.J., Delabouglise, D., Petit, J.P. and Neel, O. (**2007**) *Corrosion Science*, **49**, 818–29.

42 Racicot, R.J., Yang, S.C. and Brown, R. (**1997**) *Materials Research Society Symposium Proceedings*, **458**, 415–20.

43 Zarras, P., Anderson, N., Webber, C., Irvin, D.J., Irvin, J.A., Guenthner, A. and Stenger-Smith, J.D. (**2003**) *Radiation Physics and Chemistry*, **68**, 387–94.

44 Spinks, G.M., Dominis, A.J., Wallace, G.G. and Tallman, D.E. (**2002**) *Journal of Solid State Electrochemistry*, **6**, 85–100.

45 McAndrew, T.P. (**1997**) *Trends in Polymer Science*, **5**, 7–11.

46 Sitaram, S.P., Stoffer, J.O. and Okeefe, T.J. (**1997**) *Journal of Coatings Technology*, **69**, 65–69.

47 Yagova, I.V., Ivanov, S.S., Bykov, I.V. and Yagov, V.V. (**1998**) *Protection of Metals*, **34**, 132–36.

48 Wessling, B. (**1998**) *Synthetic Metals*, **93**, 143–54.

49 Kinlen, P.J., Menon, V. and Ding, Y. (**1999**) *Journal of Electrochemical Society*, **146**, 3690–95.

50 Paliwoda-Porebska, G., Stratmann, M., Rohverder, M., Potje-Kamloth, K., Lu, Y., Pich, A.Z. and Adler, H.J. (**2005**) *Corrosion Science*, **47**, 3216–33.

51 Dominis, A.J., Spinks, G.M. and Wallace, G.G. (**2003**) *Progress in Organic Coatings*, **48**, 43–49.

52 Kendig, M., Hon, M. and Warren, L. (**2003**) *Progress in Organic Coatings*, **47**, 183–89.

53 Narayanan, T.S.N.S. (**2005**) *Reviews on Advanced Materials Science*, **9**, 130–77.

54 Almeida, E., Diamantino, T.C., Figueiredo, M.O. and Sa, C. (**1998**) *Surface and Coatings Technology*, **106**, 8–17.

55 Lizlous, E.A. (**1976**) *Corrosion*, **32**, 263–66.

56 Magalhaes, A.A.O., Margarit, I.C.P. and Mattos, O.R. (**2004**) *Journal of Electroanalytical Chemistry*, **572**, 433–40.

57 Treacy, G.M., Wilcox, G.D. and Richardson, M.O.W. (**1999**) *Journal of Applied Electrochemistry*, **29**, 647–54.

58 Monticelli, C., Brunoro, G., Frignani, A. and Trabanelli, G. (**1992**) *Journal of the Electrochemical Society*, **139**, 706–11.

59 Guan, H. and Buchheit, R.G. (**2004**) *Corrosion*, **60**, 284–96.

60 Iannuzzi, M., Kovac, J. and Frankel, G.S. (**2007**) *Electrochimica Acta*, **52**, 4032–42.

61 Almeida, E., Fedrizzi, L. and Diamantinio, T.C. (**1998**) *Surface and Coatings Technology*, **105**, 97–101.

62 Yang, K.H., Ger, M.D., Hwub, W.H., Sungc, Y. and Liu, Y.C. (**2007**) *Materials Chemistry and Physics*, **101**, 480–85.

63 Chong, K.Z. and Shih, T.S. (**2003**) *Materials Chemistry and Physics*, **80**, 191–200.

64 Danilidis, I., Hunter, J., Scamans, G.M. and Sykes, J.M. (**2007**) *Corrosion Science*, **49**, 1559–69.

65 Yasakau, K.A., Zheludkevich, M.L., Lamaka, S.V. and Ferreira, M.G.S. (**2006**) *Journal of Physical Chemistry B*, **110**, 5515–28.

66 Bethencourt, M., Botana, F.J., Calvino, J.J., Marcos, M. and Rodriguez-Chacon, M.A. (**1998**) *Corrosion Science*, **40**, 1803–19.

67 Falconnet, P.J. (**1993**) *Journal of Alloys and Compounds*, **192**, 114–17.

68 Pardo, A., Merino, M.C., Arrabal, R., Viejo, F. and Munoz, J.A. (**2007**) *Applied Surface Science*, **253**, 3334–44.

69 Palomino, L.E.M., Aoki, I.V. and de Melo, H.G. (**2006**) *Electrochimica Acta*, **51**, 5943–53.

70 Scholes, F.H., Soste, C., Hughes, A.E., Hardin, S.G. and Curtis, P.R. (**2006**) *Applied Surface Science*, **253**, 1770–80.

71 Decroly, A. and Petitjean, J.-P. (**2005**) *Surface and Coatings Tehnology*, **194**, 1–9.

72 Bethencourta, M., Botanaa, F.J., Canoa, M.J. and Marcos, M. (**2004**) *Applied Surface Science*, **238**, 278–81.

73 Fahrenholtz, W.G., O'Keefe, M.J., Zhou, H. and Grant, J.T. (**2002**) *Surface and Coatings Tehnology*, **155**, 208–13.

74 Hinton, B.R.W., Arnott, D.R. and Ryan, N.E. (**1986**) *Materials Forum*, **9**, 162–68.

75 Neil, W. and Garrad, C. (**1994**) *Corrosion*, **50**, 215–20.

76 Wilson, L. and Hinton, B.R.W. (**1987**) Australian Patent P10649, March **1987**.

77 Hughes, A.E., Nelson, K.J., Taylor, R.J., Hinton, B.R.W., Henderson, M.J., Wilson, L. and Nugent S.A., **1995** International Patent No WO95/08008.

78 Aramaki, K. (**2005**) *Corrosion Science*, **47**, 1285–98.

79 Aramaki, K. (**2002**) *Corrosion Science*, **44**, 1621–32.

80 Aramaki, K. (**2003**) *Corrosion Science*, **45**, 1085–101.

81 Aramaki, K. (**2003**) *Corrosion Science*, **45**, 451–64.

82 Arenas, M.A. and de Damborenea, J.J. (**2004**) *Surface and Coatings Tehnology*, **187**, 320–25.

83 Ferreira, M.G.S., Duarte, R.G., Montemorb, M.F. and Simões, A.M.P. (**2004**) *Electrochimica Acta*, **49**, 2927–35.

84 Bernal, S., Botana, F.J., Calvino, J.J., Marcos, M., Pérez-Omil, J.A. and Vidal, H. (**1995**) *Journal of Alloys and Compound*, **225**, 638–41.

85 Lu, Y.C. and Ives, M.B. (**1995**) *Corrosion Science*, **37**, 145–55.

86 Montemor, M.F., Simoes, A.M. and Carmezim, M.J. (**2007**) *Applied Surface Science*, **253**, 6922–31.

87 Dabala, M., Brunelli, K., Napolitani, E. and Magrini, M. (**2003**) *Surface and Coatings Tehnology*, **172**, 227–32.

88 Kuznetsov, Y.I. (**2001**) *Protection of Metals+*, **37**, 101–7.

89 Lamaka, S.V., Zheludkevich, M.L., Yasakau, K.A., Montemor, M.F., Cecilio, P. and Ferreira, M.G.S. (**2006**) *Electrochemistry Communications*, **8**, 421–28.

90 Zheludkevich, M.L., Yasakau, K.A., Poznyak, S.K. and Ferreira, M.G.S. (**2005**) *Corrosion Science*, **47**, 3368–83.

91 Lamaka, S.V., Zheludkevich, M.L., Yasakau, K.A., Serra, R., Poznyak, S.K. and Ferreira, M.G.S. (**2007**) *Progress in Organic Coatings*, **58**, 127–35.

92 Palanivel, V., Huang, Y. and van Ooij, W.J. (**2005**) *Progress in Organic Coatings*, **53**, 153–68.

93 Kasten, L.S., Grant, J.T., Grebasch, N., Voevodin, N., Arnold, F.E. and Donley, M.S. (**2001**) *Surface and Coatings Technology*, **140**, 11–15.

94 Garcia-Heras, M., Jimenez-Morales, A., Casal, B., Galvan, J.C., Radzki, S. and Villegas, M.A. (**2004**) *Journal of Alloys and Compounds*, **380**, 219–24.

95 Voevodin, N.N., Grebasch, N.T., Soto, W.S., Arnold, F.E. and Donley, M.S. (**2001**) *Surface and Coatings Tehnology*, **140**, 24–28.

96 Trabelsi, W., Cecilio, P., Ferreira, M.G.S., Yasakau, K., Zheludkevich, M.L. and Montemorb, M.F. (**2007**) *Progress in Organic Coatings*, **59**, 214–23.

97 Pepe, A., Aparicio, M., Dur'an, A. and Cer, S. (**2006**) *Journal of Sol-Gel Science and Technology*, **39**, 131–38.

98 Montemor, M.F. and Ferreira, M.G.S. (**2007**) *Electrochimica Acta*, **52**, 7486–7495. in press doi: 10.1016/j.electacta.2006. 12.086.

99 Cabral, A.M., Trabelsi, W., Serra, R., Montemor, M.F., Zheludkevich, M.L. and Ferreira, M.G.S. (**2006**) *Corrosion Science*, **48**, 3740–58.

100 Montemor, M.F., Trabelsi, W., Zheludevich, M. and Ferreira, M.G.S. (**2006**) *Progress in Organic Coatings*, **57**, 67–77.

101 Trabelsi, W., Triki, E., Dhouibi, L., Ferreira, M.G.S., Zheludkevich, M.L. and Montemor, M.F. (**2006**) *Surface and Coatings Tehnology*, **200**, 4240–50.

102 Trabelsi, W., Cecilio, P., Ferreira, M.G.S. and Montemor, M.F. (**2005**) *Progress in Organic Coatings*, **54**, 276–84.

103 Mascia, L., Prezzi, L., Wilcox, G.D. and Lavorgna, M. (**2006**) *Progress in Organic Coatings*, **56**, 13–22.

104 Zheludkevich, M.L., Salvado, I.M. and Ferreira, M.G.S. (**2005**) *Journal of Materials Chemistry*, **15**, 5099–111.

105 Sheffer, M., Groysman, A., Starosvetsky, D., Savchenko, N. and Mandler, D. (**2004**) *Corrosion Science*, **46**, 2975–85.

106 Vreugdenhil, A.J. and Woods, M.E. (**2005**) *Progress in Organic Coatings*, **53**, 119–25.

107 Khramov, A.N., Voevodin, N.N., Balbyshev, V.N. and Donley, M.S. (**2004**) *Thin Solid Films*, **447**, 549–57.

108 Quinet, M., Neveu, B., Moutarlier, V., Audebert, P. and Ricq, L. (**2006**) *Progress in Organic Coatings*, **58**, 1–8.

109 Jakab, M.A. and Scully, J.R. (**2005**) *Nature Materials*, **4**, 667–70.

110 Jakab, M.A., Presual-moreno, F. and Sculli, J.R. (**2005**) *Corrosion*, **61**, 246–63.

111 Presuel-Moreno, F.J., Wang, H., Jakab, M.A., Kelly, R.G. and Scully, J.R. (**2006**) *Journal of Electrochemical Society*, **153**, B486–98.

112 Kendig, M. and Kinlen, P. (**2007**) *Journal of the Electrochemical Society*, **154**, C195–C201.

113 Yasuda, M., Akao, N., Hara, N. and Sugimoto, K. (**2003**) *Journal of Electrochemical Society*, **150**, B481–87.

114 Shchukin, D.G., Zheludkevich, M. and Mohwald, H. (**2006**) *Journal of Materials Chemistry*, **16**, 4561–66.

115 Shchukin, D.G. and Mohwald, H. (**2007**) *Small*, **3**, 926–43.

116 Khramov, A.N., Voevodin, N.N., Balbyshev, V.N. and Donley, M.S. (**2004**) *Thin Solid Films*, **447–448**, 549–57.

117 Uekama, K., Hirayama, F. and Irie, T. (**1998**) *Chemical Reviews*, **98**, 2045–76.

118 Rekharsky, M.V. and Inoue, Y. (**1998**) *Chemical Reviews*, **98**, 1875–917.

119 Chen, Y., Jin, L. and Xie, Y. (**1998**) *Journal of Sol-Gel Science and Technology*, **13**, 735–38.

120 Gallardo, J., Duran, A., Garcia, I., Celis, J.P., Arenas, M.A. and Conde, A. (**2003**) *Journal of Sol-Gel Science and Technology*, **27**, 175–83.

121 Conde, A., Duran, A. and de Damborenea, J.J. (**2003**) *Progress in Organic Coatings*, **46**, 288–96.

122 Malzbender, J. and de With, G. (**2002**) *Advanced Engineering Materials*, **4**, 296–300.

123 Zheludkevich, M., Serra, R., Montemor, F., Salvado, I. and Ferreira, M. (**2006**) *Surface and Coatings Tehnology*, **200**, 3084–94.

124 Zheludkevich, M.L., Serra, R., Montemor, M.F., Yasakau, K.A., Salvado, I.M.M. and Ferreira, M.G.S. (**2005**) *Electrochimica Acta*, **51**, 208–17.

125 Zheludkevich, M.L., Serra, R., Montemor, M.F. and Ferreira, M.G.S. (**2005**) *Electrochemistry Communications*, **7**, 836–40.

126 Montemor, M.F. and Ferreira, M.G.S. (**2007**) *Electrochimica Acta*, **52**, 6976–87.

127 Pippard, D.A. (**1983**) Corrosion inhibitors, method of producing them and protective coatings containing them, US Pat. 4405493, September 20.

128 R.L. Cook, Releasable corrosion inhibitor compositions, US Pat. 6933046, August 23, **2005**.

129 Schmidt, C. (**2002**) Anti-corrosive coating including a filler with a hollow cellular structure, US Pat. 6383271, May 7, 2002.

130 Boston, S. (**2007**) *Journal of Coatings Technology*, **4**, 167–75.

131 Eckler, E.P. and Ferrara, L.M. (**1988**) Anti-corrosive protective coatings, US Pat. 4738720, April 19, 1988.

132 Buchheit, R.G., Mamidipally, S.B., Schmutz, P. and Guan, H. (**2002**) *Corrosion*, **58**, 3–14.

133 McMurray, H.N., Williams, D., Williams, G. and Worsley, D. (**2003**) *Corrosion Engineering, Science and Technology*, **38**, 112–18.

134 Bohm, S., McMurray, H.N., Powell, S.M. and Worsley, D.A. (**2001**) *Materials and Corrosion*, **52**, 896–903.

135 Leggat, R.B., Zhang, W., Buchheit, R.G. and Taylor, S.R. (**2002**) *Corrosion*, **58**, 322–28.

136 Wang, H., Presuel, F. and Kelly, R.G. (**2004**) *Electrochimica Acta*, **49**, 239–55.

137 Buchheit, R.G., Guan, H., Mahajanam, S. and Wong, F. (**2003**) *Progress in Organic Coatings*, **47**, 174–82.

138 McMurray, H.N. and Williams, G. (**2004**) *Corrosion*, **60**, 219–28.

139 Williams, G. and McMurray, H.N. (**2004**) *Electrochemical and Solid State Letters*, **7**, B13–B15.

140 Yang, H. and van Ooij, W.J. (**2004**) *Progress in Organic Coatings*, **50**, 149–61.

141 Decher, G., Hong, J.D. and Schmitt, J. (**1992**) *Thin Solid Films*, **210/211**, 831–35.

142 Clark, S.L., Handy, E.S., Rubner, M.F. and Hammond, P.T. (**1999**) *Advanced Materials*, **11**, 1031–35.

143 Shchukin, D.G., Sukhorukov, G.B. and Mohwald, H. (**2003**) *Chemistry of Materials*, **15**, 3947–50.

144 Zheludkevich, M., Shchukin, D.G., Yasakau, K.A., Mohwald, H. and Ferreira, M.G.S. (**2007**) *Chemistry of Materials*, **19**, 402–11.

145 Shchukin, D.G., Zheludkevich, M., Yasakau, K., Lamaka, S., Ferreira, M.G.S. and Mohwald, H. (**2006**) *Advanced Materials*, **18**, 1672–78.

146 Shchukin, D.G. and Möhwald, H. (**2007**) *Advanced Functional Materials*, **17**, 1451–58.

147 Shchukin, D., Sukhorukov, G.B., Price, R. and Lvov, Y. (**2005**) *Small*, **1**, 510–13.

148 Lvov, Y., Price, R., Gaber, B. and Ichinose, I. (**2002**) *Colloids and Surface: Engineering*, **198**, 375–82.

149 Kommireddy, D., Sriram, S., Lvov, Y. and Mill, D. (**2006**) *Biomaterials*, **27**, 4296–303.

150 Luca, V. and Thomson, S. (**2000**) *Journal of Materials Chemistry*, **10**, 2121–26.

151 Shchukin, D.G., Lamaka, S.V., Yasakau, K.A., Zheludkevich, M.L., Möhwald, H. and Ferreira, M.G.S. (**2008**) ''Active anticorrosion coatings with halloysite nanocontainers'', *Journal of Physical Chemistry C*, **112**(4), 958–964.

5

Self-healing Processes in Concrete

Erik Schlangen and Christopher Joseph

5.1
Introduction

Concrete is the most widely used man-made building material on the planet, and cement is used to make approximately 2.5 t (over one cubic metre) of concrete per person per year [1]. Concrete structures have been built since the discovery of Portland cement (PC) in the midst of the nineteenth century. The reaction of PC with water results in hydration products, which glue the reacting cement particles together to form a hardened cement paste. When cement and water are mixed with sand, the resulting product is called *mortar*. If the mixture also contains coarse aggregate, the resulting product is called *concrete*. It is a quasi-brittle material, strong in compression but relatively week in tension. The compressive strength of traditional concrete varies between 20 and 60 MPa. By using a low water cement ratio, improved particle packing, and special additives, high strength concrete can be produced with strength values up to 150–200 MPa. Cement-based composites have even been produced with compressive strengths up to 800 MPa.

Concrete elements loaded in bending or in tension easily crack. For this reason reinforcement is installed. Passive reinforcement is activated as soon as the concrete cracks. The formation of cracks is considered an inherent feature of reinforced concrete. It must be emphasized that in reinforced concrete structures, cracks as such are not considered as damage or failure and cracking as such does not indicate a safety problem. The crack width, however, should not exceed a prescribed crack width limit. Cracks that are too wide may reduce the capacity of the concrete to protect the reinforcing steel against corrosion. Corrosion of reinforcing steel is the major reason for premature failure of concrete structures. Apart from these macrocracks, very fine cracks, that is, microcracks, may occur within the matrix due to restraint of shrinkage deformations of the cement paste. Microcracks are an almost unavoidable feature of ordinary concrete. If microcracks form a continuous network of cracks they may substantially contribute to the permeability of the concrete, thereby reducing the concrete's resistance against ingress of aggressive substances.

Self-healing Materials: Fundamentals, Design Strategies, and Applications. Edited by Swapan Kumar Ghosh
Copyright © 2009 WILEY-VCH Verlag GmbH & Co. KGaA, Weinheim
ISBN: 978-3-527-31829-2

Even though cracks can be judged as an inherent feature of reinforced concrete and the existence of cracks does not necessarily indicate a safety problem, cracks are generally considered undesirable for several reasons. The presence of cracks may reduce the durability of concrete structures. In cases where structures have to fulfill a retaining function, cracks may jeopardize the tightness of the structure. Completely tight concrete may be required in case the structure has to protect the environment against radiation from radioactive materials or radioactive waste. Cracks may also be undesirable for aesthetic reasons. Not only cracks but also the inherent porous structure of concrete can be a point of concern. If pores are connected and form a continuous network, harmful substances may penetrate the concrete and may chemically or physically attack the concrete or the embedded steel.

The performance of structures with elapse of time is often presented with graphs, as shown in Figure 5.1 [2]. Curve A describes how, after some time, gradual degradation occurs until the moment that first repair is urgently needed. The durability of concrete repairs is often a point of concern. Very often a second repair is necessary only 10–15 years later. Spending more money initially in order to ensure a higher quality often pays off. The maintenance-free period will be longer and the first major repair work can often be postponed for many years (See line B in Figure 5.1a and 5.1b). Apart from saving direct costs for maintenance and repair, the savings due to reduction of the indirect costs are generally most welcomed by the owner.

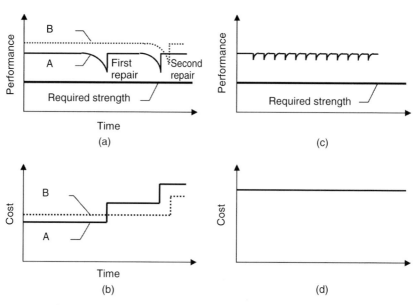

Fig. 5.1 Performance (a) and costs (b) with elapse of time for normal (A) and high quality (B) structures. Direct costs of repair included. Interest and inflation not considered. Performance (c) and cost (d) of a structure made with self-healing material (concrete) with elapse of time. Interest and inflation not considered.

The experience is that a initial higher quality of the material results in postponement of repair and, consequently, a reduction of the costs for maintenance and repair. This raises a logical question concerning the optimum balance between increasing the initial costs and the cost savings for maintenance and repair. The extreme case would be that no costs for maintenance and repair have to be considered at all because the material is able to repair itself.

Figure 5.1c schematically shows the performance of a structure made with self-repairing material. On the occurrence of a small crack or the start of any physical or chemical degradation process, the material gradually starts to repair itself and the structure will regain its original level of performance or a level close to that. Figure 5.1d illustrates the anticipated costs for such a material. In this figure, inflation and interest are not considered. The initial costs will be substantially higher than that of a structure made with traditional concrete mixtures. The absence of maintenance and repair costs, however, could finally result in a financially positive situation for the owner.

An enhanced service life of concrete structures will reduce the demand for new structures. This, in turn, results in the use of less raw materials and an associated reduction in pollution, energy consumption, and CO_2 production. Statistics convincingly show the enormous amounts of money spent by society owing to the lack of quality and durability of concrete structures. The cost for reconstruction of bridges in the United States has been estimated between $20 and $200 billion [3]. The average annual maintenance cost for bridges in that country is estimated at $5.2 billion. Comprehensive life cycle analyses indicate that the indirect costs due to traffic jams and associated lost productivity are more than 10 times the direct cost of maintenance and repair [4]. In The Netherlands one-third of the annual budget for large civil engineering works is spent on inspection, monitoring, maintenance, upgrading, and repair. In the United Kingdom repair and maintenance costs account for over 45% of the United Kingdom's annual expenditure on construction [5]. According to the Department for environment, food and rural affairs (DEFRA) [6], about half of CO_2 emissions in the United Kingdom come from buildings, of which the actual structure is a large contibutor. The production of 1 t of PC, for example, produces approximately 1 t of CO_2 when the emissions due to calcination and fuel combustion required to power the kiln are combined [1]. Considering that about 2.35×10^9 t of cement are used annually worldwide, the CO_2 emissions associated with the production of cement are very significant, and are estimated to be in the order of 5–7% of the total CO_2 production in the world. Given the rapid growth of China and India's economy, which are the two largest consumers of cement, this figure is expected to increase if the technology to produce cement remains unchanged.

Enhancing the longevity of our built infrastructure will undoubtedly reduce the impact of mankind's activities on the stability of the biosphere. Attempts to justify fundamental and risky research by promising the solutions of the serious environmental problems faced by society today may sound a bit modish. This, however, should not keep us from starting this research. Not starting this research will worsen the situation. Moreover, a direct benefit for many parties involved in the building industry and for society as a whole is conceivable if it were possible to improve the quality and service life of concrete structures. Historically, the concrete in these structures has been designed to meet predefined specifications at the start of its life. Longevity of the structure is then monitored through maintenance programs. More recently, however, material scientists have begun to adopt a change in philosophy whereby adaptability or "self-healing" of the material over time is explicitly considered. The inspiration for this change has often come from nature through biomimicry. In concrete, early research in this self-healing area has focused on both the natural ability of hydrates to heal cracks over time, and artificial means of crack repair from adhesive reservoirs embedded within the matrix.

This chapter focuses on work that has been completed recently at Delft and Cardiff Universities in both "natural" (autogenic) and "artificial" (autonomic) healing of concrete. Before this, a brief overview of the current state of the art in this area is given, in addition to an explanation of some commonly used definitions.

5.2
State of the Art

5.2.1
Definition of Terms

A number of terms, such as intelligent materials, smart materials, smart structures, and sensory structures have been used in the literature for this new field of self-healing materials. There is some ambiguity regarding the definition of these terms within the literature. However, Mihashi *et al.* [7] and Sharp and Clemeña [8] give some insight into their meaning.

5.2.1.1 Intelligent Materials
Intelligent materials, as defined by Mihashi *et al.* [7], are materials that "incorporate the notion of information as well as physical index such as strength and durability". This higher level function or "intelligence" is achieved through the systematic corporation of various individual functions. As a result, intelligent materials exhibit a self-control capability whereby they are not only able to sense and respond to various external stimuli but also to conduct this response in a regulated manner. This is analogous to the behavior exhibited by many natural materials such as skin, bone, and tendons. Schmets [9] identifies this inherent "intelligent" adaptability of natural materials, and states that their outstanding mechanical properties are a consequence of their highly organized hierarchical structure, which is omnipresent at all levels (length scales) of the material.

Given their complexity, it is not surprising that such materials are currently not used in practice. The development of man-made intelligent materials are still largely at the conceptual and early design stages, and are confined mainly to the fields of medicine, bionics, and aeronautics/astronautics [7].

5.2.1.2 Smart Materials

Smart materials are engineered materials that are able to provide a unique beneficial response when a particular change occurs in its surrounding environment [8]. Examples of smart materials include piezoelectric materials, magnetostrictive materials, shape memory materials, temperature-responsive polymers that are able to change color with temperature, and smart gels that are able to shrink or swell by factors of up to 1000 in response to chemical or physical stimuli. The difference between a smart material and an intelligent material is therefore defined by the degree to which the material can gather information, process this information, and react accordingly.

5.2.1.3 Smart Structures

Smart structures differ from smart materials in that they are engineered composites of conventional materials, which exhibit sensing and actuation properties, due to the properties of the individual components.

Self-healing Materials   Many self-healing materials fall into the category of smart structures, since they contain encapsulated healing agents that are released when damage occurs, thereby "healing" the "injury" and increasing the materials' functional life.

Self-healing studies have been performed on polymers, coatings, and composites (including concrete); however, all these "structures" rely on previous knowledge of the damage mechanisms to which they are susceptible, and are therefore classed as smart rather than intelligent.

Autonomic and Autogenic Healing   A composite material which exhibits self-healing capabilities due to the release of encapsulated resins or glues as a result of cracking from the onset of damage is categorized as having autonomic healing properties (Figure 5.22). Recent research on the autonomic healing of cementitious materials is discussed in Sections 5.2.3 and 5.4.

If the healing properties of a material are generic to that material, then the material could potentially be classed as a smart material, and the healing process is termed *autogenic healing*. Cementitious materials have this innate autogenic ability to self-repair, since rehydration of a concrete specimen in water can serve to kick-start the hydration process, when the water reacts with pockets of unhydrated cement in the matrix. Further discussion on autogenic healing of concrete is given in Sections 5.2.2 and 5.3.

Passive and Active Modes   Smart self-healing structures may also be classified depending on the passive or active nature of their healing abilities. A passive mode smart structure has the ability to react to an external stimulus without the need for human intervention, whereas an active mode smart structure requires intervention in order to complete the healing process. Both the systems have been tested, with respect to concrete by Dry [10], and are illustrated in Figures 5.2 and 5.3.

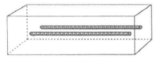

Fibers coated with wax and
filled with methylmethacrylate

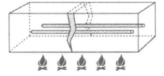

Wax coating melted with initial
heating, methylmethacrylate
released from fibers into cracks

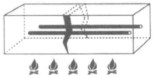

Methylmethacrylate polymerized
during second heating, closing
previous cracks

Fig. 5.2 Active release mode illustrated through the melting of a wax coating on porous fibers containing methylmethacrylate healing agent (Dry [10]).

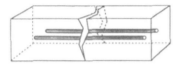

Adhesive released by cracking

Fig. 5.3 Passive release mode illustrated through the physical cracking of the brittle fiber under loading (Dry [10]).

A fully passive release system draws its main benefits from the omittance of the need for human inspection, repair, and maintenance. The requirement for human intervention in an active mode system, nonetheless, allows for a larger degree of control to be exercised, and is thus likely to inspire greater confidence within the end user.

5.2.1.4 Sensory Structures

Sensory structures are the fourth and least intelligent of the advanced materials categories. These structures have sensing capabilities but lack actuating properties. Examples of sensory structures include smart brick, which are able to monitor temperature, vibration, and movements within buildings; smart optical fibers, which are able to sense undesirable chemicals, moisture, and strain; and smart paints, which contain silicon-microsphere sensors, and are able to monitor their condition and protection effectiveness.

5.2.2
Autogenic Healing of Concrete

The autogenic healing of cementitious materials is a natural phenomenon that has been known about for many years. This effect is generally acknowledged as one of the reasons how so many old buildings and structures have survived for so long with limited servicing and maintenance. Westerbeek [11], attributes the unexpected longevity of many old bridges in Amsterdam to this effect. It is believed that this longevity is due to the high levels of chalk or calcium in the cement of that area. In the presence of water, this calcium is believed to dissolve and then deposit in cracks, thus partially healing them and hindering their propagation. Recently, the autogenic healing of microcracks has been the suggested reason for the reduction in diffusion coefficient of concrete marine structures with time. However, the durability benefit that the self-healing of microcracks afford is reduced for non-submerged concrete structures. Periodic "wetting" of structures may improve this situation; however, such a process is expensive and generally impracticable for most situations.

As described by Schlangen [12], one of the first works on the healing of cementitious materials was published more than 20 years ago. It was shown that the so-called Kaiser effect (absence of acoustic emission, which is usually observed at repeated loading of structural elements, until the load exceeds the previously achieved level) disappeared for concrete that had been kept in water for a long period before reloading.

The primary causes of autogenic self-healing are considered to be based on chemical, physical, and mechanical processes. The main processes are (i) swelling and hydration of cement pastes, (ii) precipitation of calcium carbonate crystals, and (iii) blockage of flow paths due to deposition of water impurities or movement of concrete fragments that detach during the cracking process [13].

Many authors have investigated this phenomenon in recent years: Zhong and Yao [14] have investigated the effect of the degree of damage on self-healing ability of normal strength and high strength concrete; Reinhartd and Jooss [15] have examined the effect of temperature on permeability and self-healing of cracked concrete; Şahmaran and Li [16] have considered the effect of autogenous healing on engineered cementitious composites (ECCs); and Jacobsen and Sellevold [17] have examined the efficacy of autogenic healing on strength recovery of *well cured* concrete beams exposed to rapid freeze/thaw cycles. The latter paper concluded that only a 4–5% recovery of compressive strength by means of autogenous healing was possible. However, as noted by Edvardsen [18], the greatest potential for autogenous healing exists in early age concrete. More recent work has, therefore, focused on examining not only the permeability reduction in early age cracked concrete due to autogenous healing but also the mechanical strength gain [19]. Further details of this work are given in Section 5.3.

An interesting recent development has been the autogenous healing of expansive concretes as studied by Japanese researchers ([13, 20, 21]). They have found, through microscopic observations and subsequent water permeability tests, that the inclusion of expansive agents in the concrete has allowed even large cracks of up to 0.3–0.4 mm to heal [20]. The authors have also found that the addition of small amounts of various carbonates such as bicarbonate of soda increases the self-healing ability of the concrete by allowing more calcium carbonate ($CaCO_3$) to be precipitated [21].

5.2.3
Autonomic Healing of Concrete

The autonomic healing of cementitious composites has received significantly less attention than the autogenic healing process described above, and the research efforts to date have been far more piecewise and sporadic. Nevertheless, several authors (Dry [10, 22]; Dry and Corsaw [23]; Dry and Unzicker [24]; Li *et al.* [25]; Mihashi *et al.* [7]) have undertaken studies in this area.

5.2.3.1 Healing Agents

Various healing agents have been proposed in the studies that have been undertaken on the self-healing of concrete. In contrast to the specialist healing agents employed in polymers, these healing agents have generally been "off the shelf" agents. The relatively "low cost" and readily available nature of "off the shelf" products are important assets that must be possessed by any healing agent proposed for application to a large bulk material, such as concrete.

The main healing materials that have been proposed to date are epoxy resins, cyanoacrylates, and alkali–silica solutions. It is obvious that the effectiveness of the healing process is not only dependent on the capillary forces that are dictated by the crack width, but also on the viscosity of the repair agent; the lower the viscosity the larger the potential repair area. Another prerequisite for the agent is that it must form a sufficiently strong bond between the surfaces of the crack, in order to prevent the reopening of the crack, and thus force other new cracks to open. This increases the total fracture energy that is required to break the specimen.

Epoxy Resins Low-viscosity epoxy resins currently form the principle healing agent used in the postdamage "active" remediation of critical concrete floors, and bridge decks. Sikadur 52 (Sika Ltd) and Tecroc Epoxy Injection Grout (Tecroc Products Ltd) have viscosities at room temperature (20 °C) in the order of 500 and 200 cPs, respectively. (Note: water has a viscosity of 1 cPs.(1 mPa · s), milk has 3 cPs, and grade 10 light oil has 85–140 cPs.) Epoxy resins are durable materials that generally have good thermal, moisture, and light resistance. They are available in either one or two-part systems: a one-part epoxy is activated by the presence of heat and a two-part epoxy is cured by the presence of both a hardener and resin component.

Nishiwaki *et al.* [26] have recently designed and developed a single-agent epoxy resin-based self-healing system for concrete. The low-viscosity epoxy resin is stored in an organic (ethylene vinyl acetate) film pipe that melts at 93 °C. The authors have managed to maintain a passive system by embedding a "self-diagnosis composite" sensor adjacent to the repair agent supply tube in the concrete. This sophisticated self-diagnosis composite sensor is made from a fiber-reinforced composite and an electroconductive material. When a crack forms, the sensor detects the increase in strain through reduced electrical conductivity, and as a result of the corresponding increase in resistance, heat is generated, which melts the organic supply tube and cures the epoxy resin after it has flowed into the crack.

The main problem with the application of two-part epoxy resins to the autonomic healing of concrete is that both the components have to be simultaneously present at a crack location. Given that both liquids must be encapsulated, the likelihood of both the capsules being present at a crack location, and cracking at the same time, is extremely small. Mihashi *et al.* [7] tried to overcome this by manually placing the two components in adjacent tubes. Despite both the tubes cracking and releasing their respective agents, poor mechanical behavior was observed due to insufficient mixing of the fluid blend.

The chemical reaction that occurs between the hardener and the resin is an exothermic reaction, which, unfortunately, does not rely on the presence of oxygen to perpetuate. An encapsulated mixture of both the agents will, therefore, only remain liquid for the duration of the "pot life" of the resin, which is usually in the order of hours.

Cyanoacrylates Cyanoacrylates (superglues) are one-part systems that react to the presence of moisture, and are noted for their ability to cure rapidly (pot life in the order of seconds–minutes). They provide a bond strength that often exceeds the strength of the substrate, certainly in the case of concrete. They also have very low viscosities (<10 cPs), and therefore possess the ability to heal cracks less than 100 µm in thickness. Li et al. [25] studied the effectiveness of an ethyl cyanoacrylate as a healing agent in ECCs, and observed an increase in initial stiffness in most specimens.

An important property of cyanoacrylates in relation to their use in concrete is the fact that they are acidic solutions. Contact with concrete, which is an alkaline material (pH of approximately 13), will result in neutralization of the glue and thus quicker setting times. This quicker gain of bond strength can be beneficial in rapid cyclic loading conditions; however, if the setting time is too quick, the dispersion of the healing agent within the crack may be insufficient.

Alkali–silica Solution Mihashi et al. [7] present the use of a diluted and undiluted alkali–silica solution as a healing agent in concrete. The alkali–silica solution in the presence of oxygen causes hydration, thereby bonding the original crack faces together. The strength of the bond is less than that of glue, although this is immaterial as long as the bond strength is greater than the tensile strength of the surrounding material. The use of an alkali–silica-based healing material in concrete is also likely to cause less material compatibility problems than its polymer-based counterparts.

5.2.3.2 Encapsulation Techniques

Various encapsulation techniques have been proposed in the literature. Mihashi et al. [7] discuss two encapsulation techniques, namely, a microcapsule enclosing repairing agent mixed in concrete and a continuous glass supply pipe enclosing repairing agent embedded in concrete, while Li et al. [25] used cyanoacrylycate enclosed in capillary tubes sealed with silicon.

The shape of the embedded capsule is a factor that should be considered. A spherically shaped capsule will provide a more controlled and enhanced release of the healing agent upon breakage, and will also reduce the stress concentrations around the void left from the empty capsule. A tubular capsule, however, will cover a larger internal area of influence on the concrete for the same volume of healing agent (higher surface area to volume ratio). The potential release of the healing agent upon cracking, however, will be reduced since localized and multiple cracking may occur, thus inhibiting the effective distribution of the healing agent.

Microcapsules The advantage of dispersed microcapsule inclusion is that the concrete can react to diffuse cracking at multiple locations, and the disadvantage is that additional repairing agent cannot be supplied once the original agent has been exhausted.

In preliminary investigations by Mihashi *et al.* [7], urea-formaldehyde microcapsules (diameter 20–70 µm) filled with epoxy resin and gelatin microcapsules (diameter 125–297 µm) filled with acrylic resin were used in compression and splitting tests. In addition to the aforementioned problems of blending two-agent epoxies, the authors concluded that the quantity of repairing agent provided by microencapsulation is very small and limited, and the bond strength between the microcapsule and the matrix needs to be stronger than the strength of the microcapsule.

Glass Capillary Tubes Li *et al.* [25] utilized capillary tubes, developed for blood testing in the medical industry, as encapsulating containers for an ethyl cyanoacrylate healing agent. Their initial test regime was aimed at confirming the sensing and actuation mechanisms of an ECC, through forced cracking of single hollow capillary tubes, under the eye of an environmental scanning electron microscope (ESEM). Custom made hollow capillary tubes, 500 µm in diameter and 60 µm wall thickness, were used, as illustrated in Figure 5.4 below.

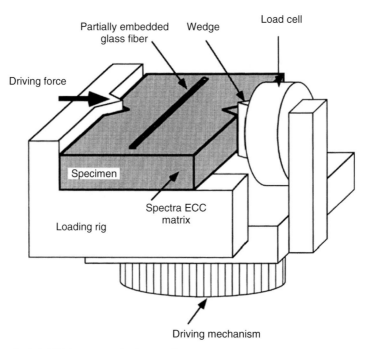

Fig. 5.4 ESEM specimen loading configuration (specimen dimensions are 10 mm × 10 mm × 1.5 mm) (Li *et al.* [25]).

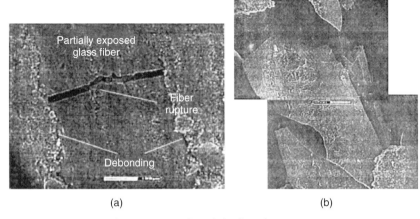

(a)   (b)

Fig. 5.5 ESEM images showing (a) tensile and (b) flexural failure of hollow capillary tubes (Li *et al.* [25]).

It was observed in the experiments that as the angle between the crack plane and the longitudinal direction of the capillary tube became more acute, the failure of the tube changed from a simple tensile mode (Figure 5.5a) to a flexural mode (Figure 5.5b). Some localized debonding between the borosilicate tube and cement matrix was also observed; however, subsequent testing of dye-filled tubes illustrated the success of the sensing and actuation mechanisms.

Continuous Glass Supply Pipes   In later experimental work conducted by Mihashi *et al.* [7], the authors reverted to the use of glass supply pipes as oppose to individual microcapsules or capillary tubes. In early work presented by Dry [10], the active provision of healing agent was proposed by the use of an internal delivery vacuum pressure system, as shown in Figure 5.6.

Continuous glass supply pipes, with or without vacuum pumps, have the advantage of being able to allow the type of repairing agent to be varied, and additional supply to be provided. This allows larger fractures to be healed compared with other encapsulation methods. The significant disadvantage of the method is the care that must be taken during casting to avoid tube breakages. Hence, this method is not suitable for *in situ* casting of concrete. It does, however, provide an interesting feasibility test for the concept of self-healing in cementitious materials.

Replacing adhesive by vaccuum

Fig. 5.6 Design for the interior delivery of chemicals from fibers by vacuum pressure (Dry [10]).

An alternative, although more complex, solution is to use organic flexible supply tubes as discussed previously in Section 5.2.3.1.

5.3
Self-healing Research at Delft

5.3.1
Introduction

The study of self-healing concrete started in Delft in 2005 with a study on the crack healing in early age concrete. In massive concrete structures, such as tunnels or bridges, often surface cracks develop during the first day of hydration [27, 28]. With continuation of the hydration process and cooling down of the interior of the structure, the surface is loaded in compression. The cracks are closed due to this compressive stress. The question is whether the cracks can heal in this situation? Hydration is still in progress and the crack faces are touching each other again. Therefore, an environment is created in which the cracks may potentially heal. To study this, a combined experimental and numerical investigation was started, which is described in the following paragraphs.

In 2005, a big initiative was also started by the Delft Centre for Materials (DCMat) [29] at Delft University of Technology, to study self-healing mechanisms in various materials. A few examples of the projects started are "nanoscale characterization of self-healing in aluminium based alloys", "self-healing intermetallic-(metal, polymer) matrix composites", "self-healing elastomer nanocomposites", and "blood vessels in polymers for self-healing". A description of all the projects can be found in Ref. 29. As part of this initiative of DCMat, at the Civil Engineering Department a project was set up to investigate the self-healing of concrete by bacterial mineral precipitation [30]. The initial results of this study are discussed in this chapter. In 2007, a large national project was initiated in The Netherlands where a large number of studies were started, including several on cement-based materials, which are all still running [31]. Some insight into these studies is given in the last part of this chapter with a view to future work in this area.

5.3.2
Description of Test Setup for Healing of Early Age Cracks

In order to study the effect of crack healing in early age concrete, first cracks have to be made in the concrete in a controlled way. For this purpose, a three-point bending test was chosen on prismatic concrete specimens. The specimens were cast in steel moulds and had dimensions of $40 \times 40 \times 160\,mm^3$. The distance between the loading points was 105 mm (Figure 5.7). The reaction force was in the center. The deformation was measured with two Linear variable differential transformers (LVDTs) (at front and back) fixed at the bottom of the specimen (in the center). The measuring length was 55 mm. The deformation measured with these LVDTs gives a bending strain at the bottom of the specimen. If the crack is

Fig. 5.7 Experimental setup.

localized, this value is a measure (although not exact) of the crack opening. The specimen was loaded in a special three-point bending frame (Figure 5.7) in an Instron 8872 servo-hydraulic loading device.

The tests were performed in deformation control using the average signal of the two LVDTs as the feedback signal. In the test, the loading was stopped at a crack mouth opening of 50 μm. The specimen was unloaded and taken out of the machine. The crack opening decreases after unloading but the crack will not completely close. At this moment, the specimen was placed in a compression loading device that applies a compressive force to close the crack. This force was measured by means of the deformation of a calibrated steel spring. The amount of compressive force was varied in the tests. In Figure 5.8, the compressive loading devices are shown with specimens subjected to 0.5, 1, and 2 MPa compressive stresses. The specimens were then stored for a specific period at a certain temperature and relative humidity (RH) (or in water) to undergo crack healing. In addition to healing of the crack, the concrete will also undergo further hydration, which means that the mechanical properties of the material itself will also improve.

In order to examine the mechanical properties of the healed crack, the specimens were, after a certain period of healing, again tested in three-point bending. To be able to judge the recovery of the mechanical properties of the crack, the results have to be compared with cracks that have not healed and with cracks that have been

Fig. 5.8 Compressive loading devices.

made after the healing process. Therefore, virgin specimens stored in the same environment and with the same age as the healed specimens were also tested. First, these specimens are loaded in deformation control up to a displacement of 50 μm. Then, the specimen is unloaded and tested again. The maximum load reached in the second stage of the test is taken to be the maximum flexural stress for a specimen with an unhealed crack at an age of two weeks.

A question that emanates from this work is whether the crack has the same length in a specimen tested at an age of one day as at two weeks, when in both cases the crack opening has reached a value of 50 μm. To answer this question, the specimens have been vacuum impregnated with a fluorescence epoxy after the test. The cracks were then visualized under UV-light. The scatter in the crack length that was observed is rather large, but it turned out that in all the situations at a crack width of 50 μm, the crack tip had reached about half way up the height of the specimen.

5.3.3
Description of Tested Variables

In the experiments, several parameters were varied. The parameters discussed in this chapter are as follows:

- The amount of compressive force applied during healing.
  The variation of this parameter is 0.0, 0.5, 1.0, and 2.0 MPa.

Table 5.1 Mixture composition of self-healing concrete specimens.

Cement	CEM III/b 42,5 LH HS	$375 \, \text{kg m}^{-3}$
Water		$187.5 \, \text{kg m}^{-3}$
River gravel	8–4 mm	$540 \, \text{kg m}^{-3}$
	2–4 mm	$363 \, \text{kg m}^{-3}$
	1–2 mm	$272 \, \text{kg m}^{-3}$
	0.5–1 mm	$272 \, \text{kg m}^{-3}$
	0.25–0.5 mm	$234 \, \text{kg m}^{-3}$
	0.125–0.25 mm	$127 \, \text{kg m}^{-3}$

- The type of cement on the concrete mix, both a blast furnace slag cement (BFSC) and an ordinary portland cement (OPC) are tested.
- The moment of creating the first crack in the specimen. The cracks are made at an age of 20, 27.5 (further named one day), 48, and 72 h. The age at loading has some variation, since the specimens are cast at the same time, but for testing only one machine is available. Each test, including preparation, takes about 45 min.
- The crack (mouth) opening of the crack. Initial crack openings of 20, 50, 100, and 150 μm are discussed in this chapter.
- The influence of the RH during healing. Specimens are stored in water and in a climate chamber with 95% and 60% RH, respectively.

In all cases, only one parameter is varied at any one time. The default parameters in the tests are 1.0 MPa compressive stress, concrete made with BFSC, crack made at age of one day, crack opening of 50 μm, and healing in water. All the tests are performed at least three times. In Table 5.1 the mix composition of the concrete is given, where the concrete mix is changed, and the CEM III (BFSC) is replaced by the same amount of CEM I 52,5 R (OPC).

5.3.4
Experimental Findings

In this section, the results of the experimental findings are briefly described. More detailed results can be found in Refs [19, 32].

5.3.4.1 Influence of Compressive Stress
The first parameter investigated is the compressive stress on the specimen during crack healing. The specimens were tested at an age of one day up to a crack opening of 50 μm. Then a compressive force was applied to the specimens and they were stored for two weeks in water and tested again. In Figure 5.9 the flexural stress is plotted versus displacement for the reloading test after two weeks of healing for a specimen with (1 MPa) and without

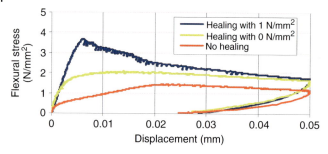

Fig. 5.9 Flexural stress against displacement for two-week-old specimens.

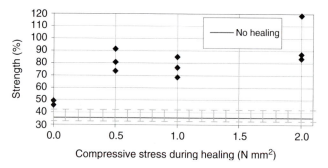

Fig. 5.10 Relative strength of specimens after healing.

(0 MPa) applied compressive stress. Furthermore, the graph shows the specimen without healing. The latter is obtained after reloading a specimen tested at an age of two weeks. The graph shows that when the crack is not closed (the compressive stress is $0\,N\,mm^{-2}$), the recovery of strength is minor. The same observations were recently done for similar tests on High performance concrete (HPC) [33]. However, with a compressive stress of 1 MPa, both stiffness and strength of the specimen is recovered and show values close to the reference specimen.

Figure 5.10 shows the relative strength of the specimen after healing for different amounts of compressive stress applied during healing. The relative strength is given as a percentage of the strength of the uncracked virgin specimen tested at an age of two weeks. In the figure, a line is also shown (with vertical bars showing the scatter), which represents the strength of the material of the unhealed specimen. The figure clearly shows that almost no increase in strength is obtained when the specimen is not loaded in compression ($0\,N\,mm^2$). Furthermore, it can be seen that when compressive loading is applied to close the crack, the amount of compressive stress does not significantly influence the strength gain.

5.3.4.2 Influence of Cement Type
The influence of the cement type in the concrete mix is presented in Figures 5.11 and 5.12. In Figure 5.11a, the flexural stress is plotted versus the crack opening

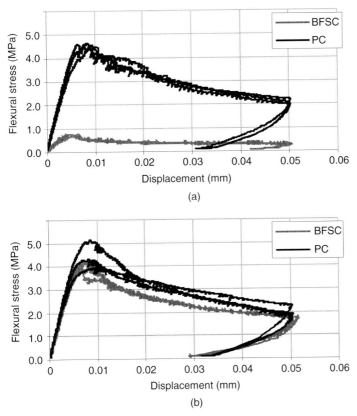

Fig. 5.11 Flexural stress versus crack opening for tests after one day (a) and after healing (b) for BFSC (blast furnace slag cement) and PC (Portland cement) concrete.

displacement for the test performed on a one-day-old specimen. It can be seen that the specimens containing the faster reacting PC have much higher strength gain. The concrete with PC is further matured. In Figure 5.11b the result is plotted for the tests after healing on both materials. Now the strength for both the materials is almost equal. Also, for the specimen tested for the first time after 15 days (Figure 5.12a), it can be seen that the strength of both the materials is almost equal. Figure 5.12b shows the plots for the reloading test, which represents the strength of the unhealed specimens. From these tests, it is clear that in the BFSC-concrete the strength gain after one day is minimal. Healing of the crack can then take place more readily because there is still a large amount of hydration of the cement, which has not yet taken place. However, in the PC-concrete the result obtained is remarkable, since here also the strength is almost fully recovered. Probably this concrete has a large amount of unhydrated cement left. Thus, the potential for crack healing is much higher in PC-concrete.

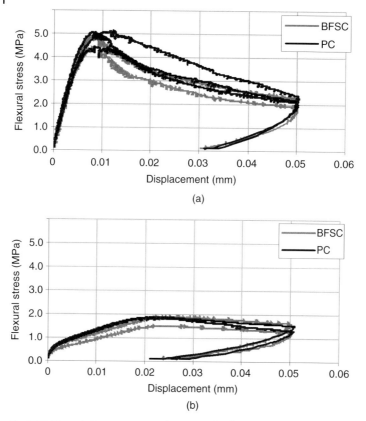

Fig. 5.12 Flexural stress versus crack opening for first test
(a) and reloading (b) after 15 days for BFSC (blast furnace
slag cement) and PC (Portland cement) concrete.

5.3.4.3 Influence of Age When the First Crack is Produced

The third parameter investigated is the moment of cracking or the age of the
specimen when the first crack is produced. In these tests, the crack is opened
up to a crack mouth opening of 50 μm. Subsequently, the specimens are loaded
in compression with a compressive stress of 1 MPa and stored in water for two
weeks. The strength after healing (relative to the strength of the virgin specimen)
is plotted in Figure 5.13 for the various ages at which the first crack is made. The
reference test for each series is always performed at the same age. So, this means
that, for instance, the strength of the specimen tested for the first time at 72 h and
subsequently healed for two weeks is compared with the strength of a specimen
loaded for the first time at an age of 17 days. The difference in strength of the virgin
material between the ages of 14 and 17 days is, however, very low. A clear decrease
in strength recovery is observed with increasing age of the specimen when the first
crack is made.

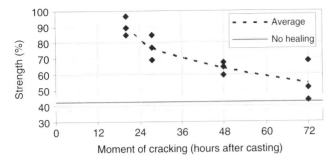

Fig. 5.13 Relative strength for specimens tested at different ages.

5.3.4.4 Influence of Crack Width

The fourth parameter discussed in this section is the influence of the width of the crack that is made in the specimen on the strength recovery. In these tests, the specimens are loaded at an age of one day and the compressive stress during healing is equal to 1.0 MPa. A larger crack mouth opening will result in a longer crack, which has propagated further into the specimen. The load that can be carried at a larger crack opening will be smaller. The results show a significant amount of scatter, but there seems to be no influence of crack opening on the strength recovery after healing.

5.3.4.5 Influence of Relative Humidity

The last parameter studied is the influence of the RH on crack healing. It turned out that only for the case when the specimens were stored in water during the healing period, recovery of strength was possible. For the cases when the specimens were stored in an environment of 95 or 60% RH, almost no increase in strength was observed. For the case of 95% RH, the specimens were even stored for a period of three months; however, even under these conditions, no crack healing was observed.

5.3.5
Simulation of Crack Healing

For the simulation of crack healing, the module MLS of FEMMASSE [34] was used. MLS is a finite element model based on the state parameter concept. This means that the material properties are a function of the state of the material. The state can be maturity, degree of hydration, temperature, or moisture potential. The simulations with the model were performed in order to check the hypothesis that the healing observed in the experiments is due to ongoing hydration of the cement. The ageing of the concrete and the development of the properties in time were taken into account in the analysis. In the analysis, a concrete with BFSC was used. The purpose of the analysis was not to exactly match the behavior as found in the experiment, but more to be able to simulate the mechanism. The properties of the material used in the simulation correspond with an average

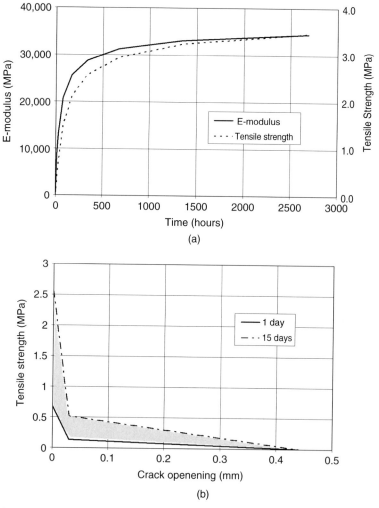

Fig. 5.14 (a) Development of E-modulus and tensile strength in time and (b) softening behavior in interface element.

concrete with BFSC, a water cement ratio of 0.5, and a compressive strength of about 45 MPa. The development of the E-modulus and the tensile strength is shown in Figure 5.14.

In the analysis, the three-point bending test was simulated in 2D using plane stress conditions. In the center, under the loading point, a discrete crack was created in the mesh using an interface element with a stress–crack opening relation. The strength of the interface element was taken to be equal to the strength of the neighboring concrete at the specific age when the three-point bending test was performed. The values used for the different simulations are shown in Table 5.2.

Table 5.2 Values used for different simulations.

Simulation	Tensile strength (MPa)
1 d	0.67
15 d	2.56
1–15 d	1.89
1–final	2.79
(Final strength)	(3.46)

The softening relation was taken to be bilinear in all cases, with the bending point in the descending branch set at 20% of the strength and at a crack opening of 0.03 mm. The final crack opening (at zero load) was set to 0.44 mm. In Figure 5.14b, the bilinear softening curve for a concrete at 1 and 15 days old is shown. Also, a simulation was performed of a specimen that was cracked at an age of 1 day and then healed for a period of 14 days. It is assumed that the bottom half of the specimen is cracked and that the strength gain of the material in the crack is equal to the strength gain of the concrete itself in these 14 days. This means that the strength of the interface is equal to the strength at 15 days minus the strength at 1 day. The softening relation is then equal to the gray area on the graph in Figure 5.14b. In the experiments, the cracked specimens were stored in water during healing. This means there is sufficient water available for the crack to heal. The unhydrated cement particles present in the crack are, therefore, possibly able to fully hydrate and are thus able to generate a higher strength than similar particles situated in the bulk of the specimen. To check the maximum load that could be reached in such a case, a simulation is performed in which the strength of the interface is taken as equal to the final strength of the specimen after full hydration minus the strength at one day.

The results of the simulations are shown in Figures 5.15–5.17. Figure 5.15 shows a typical deformed mesh in which the open discrete crack is visible. Figure 5.16 shows a stress-contour plot when the crack tip is at half the height of the specimen. It can also be seen from this figure that the crack is not completely stress

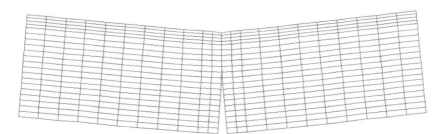

Fig. 5.15 Typical deformed mesh.

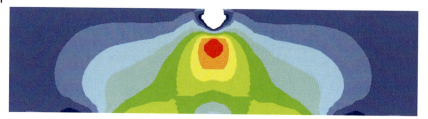

Fig. 5.16 Stress contour plot.

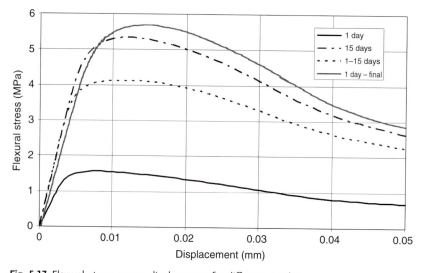

Fig. 5.17 Flexural stress versus displacement for different simulations.

free, but stress can be transferred by the interface when it is in the descending branch.

In Figure 5.17, the flexural stress is plotted versus the displacement (same measuring length as in the experiments (55 mm)) for the different simulations. From this graph, the following observations can be made:

- The flexural stress as well as the stiffness in the specimen tested at one day is lower than the specimen tested at 15 days.
- The specimen that is precracked after 1 day and tested after 14 days of healing (1–15 days) has the same stiffness as the uncracked specimen tested at 15 days. The strength of this specimen is about 77% of the uncracked specimen.
- If the strength gain in the 14 days of healing is equal to the final strength minus the strength at 1 day, then the obtained flexural stress is higher (about 106%) than the strength of the uncracked specimen tested at 15 days.

5.3.6
Discussion on Early Age Crack Healing

This section discusses the outcome of a combined experimental and numerical investigation performed at the Microlab of Delft University on crack healing of cracks in early age concrete. Prismatic specimens at a certain age are cracked in three-point bending after which they are subjected to a compressive load and stored in water to heal.

The variables investigated and discussed in this section are as follows:
- amount of compressive stress during healing;
- type of cement in the concrete;
- age of specimen when making the first crack;
- crack mouth opening of the crack;
- RH during the healing.

From the experimental results it can be concluded as follows:
- Cracks do heal when the conditions are such that the cracks are made at an early age and the cracks are closed again (a compressive stress is applied) and the specimens are stored in water.
- The amount of compressive stress does not seem to influence the strength recovery. The results indicate that a compressive stress is needed to close the crack, but once the two crack faces touch each other, or the distance between the crack faces is small enough, crack healing can occur.
- For concrete made with both BFSC and OPC, crack healing takes place. Most probably in the case of OPC there is a lot of unhydrated cement left in the crack. Storing the specimens in water probably opens the way to further hydrate this cement in the crack. In the bulk material, water cannot reach these unhydrated particles. This means that concrete with OPC probably has additional capacity for crack healing at a later stage when compared to BFSC-concrete.
- With increasing age of the specimen from the moment the first crack is made, a decrease in strength recovery is found. The age of the specimen when the first crack is made indicates the degree of hydration that has already occurred. The amount of hydration that can still take place is therefore fixed. The amount of potential strength recovery is therefore limited when the concrete has already reached a high degree of hydration when the crack is made.
- The width of the crack does not seem to influence the strength recovery due to healing. The tests with different crack mouth openings show similar amounts of strength recovery.

- Crack healing is only observed when the cracked specimens are stored in water.

The authors believe that ongoing hydration is the mechanism for crack healing that leads to the strength recovery in this investigation. This mechanism works only when the crack is closed again. It has been shown that crack healing does take place when enough water is present. The simulations that have been performed strengthen this hypothesis. It has been shown through simulation that the increase in strength in the crack due to further hydration could be sufficient to explain the observed recovery in flexural strength found in the experiments. The simulations also showed that higher strengths can be obtained in the crack compared to the bulk material when it is assumed that owing to the water in the crack the final degree of hydration is reached faster in this zone.

For the practical situation of early age surface cracks in (massive) concrete structures, which are a concern from a durability point of view, this investigation shows some promising results. It indicates that these surface cracks can disappear again, at least under the right conditions as discussed above.

5.3.7
Measuring Permeability

In the experimental work discussed above, the focus was on measuring the mechanical properties after self-healing. A major concern in concrete structures is the effect that cracks have on durability. In this case, the most important aspect is the blocking of the transport through cracks and decreasing the permeability. To measure this decrease of permeability as a result of self-healing of cracks, a new device has been developed, similar to the machine described in Ref. 35. In this device, the permeability of a (cracked) concrete disc can be measured. To fracture the concrete disk, a deformation-controlled tensile splitting test is performed (Figure 5.18). The crack opening measured with LVDTs on the front and back of the specimen is used as a feedback signal. At a certain crack opening, the test is stopped and the specimen is unloaded, as shown in Figure 5.19. In the present research, the concrete discs are tested at the age of one day. Next, the cracked specimens are loaded in compression in the same device, as shown in Figure 5.18, to close the crack again. Subsequently, the specimens are stored in water and the permeability is tested after two weeks of healing using the device as shown in Figure 5.20. The discs are placed in a steel ring and an epoxy is used to ensure that there is no leakage at the edges. The steel ring is placed in the device and a water pressure (between 1 and 6 bar, depending on the permeability of the sample) is applied at the top of the specimen. The water percolating through the specimen is measured over time. At present, the tests are still in process, but the first results look promising. Owing to the ongoing hydration, the permeability of the cracks decreases. The device will also be used to study the decrease of permeability in other ongoing self-healing studies at the Microlab (i.e. the Bacterial concrete as described in the following paragraph).

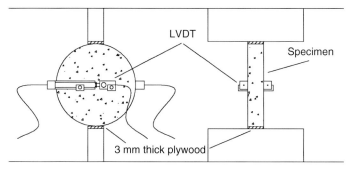

Fig. 5.18 Tensile splitting test on concrete discs.

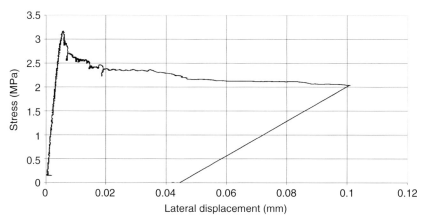

Fig. 5.19 Typical stress—crack opening displacement in tensile splitting test.

5.3.8
Self-healing of Cracked Concrete: A Bacterial Approach

Another self-healing mechanism could be based on the addition of a self-healing agent that would make up part of the concrete matrix without or insignificantly affecting its structural and mechanical characteristics. In this study, the potential of bacteria to act as a self-healing agent in concrete is investigated. Although the idea to use bacteria and integrate them in the concrete matrix may seem odd at first, it is not so from a microbiological viewpoint. Bacteria naturally occur virtually everywhere on earth, not only on the surface but also deep within, for example, in sediment and rock at a depth of more than 1 km [36]. Various species of so-called extremophilic bacteria, that is, bacteria that love the extreme, are found in highly desiccated environments such as deserts [37, 38], inside rocks [39], and even in ultrabasic environments [40, 41] that can be considered homologous to the internal concrete environment. Typical for many desiccation- and/or alkali-resistant bacterial species is their ability to form endospores. These specialized cells, which are characterized by an extremely low metabolic activity, are known to be able to resist high

Fig. 5.20 Setup for permeability measurements.

mechanically- and chemically induced stresses [42] and are viable for periods of up to 200 years [43]. In some previously published studies, the application of bacteria for cleaning of concrete surfaces [44] and improving the strength of cement–sand mortar [45] was reported. Furthermore, in some studies, the crack-healing potential by mineral-precipitating bacteria on degraded lime-stone [46] and ornamental stone surfaces [47] as well as on concrete surfaces [48, 49] was investigated and reported. Although promising results were reported, the major drawback of the latter studies was that the bacteria and compounds needed for mineral precipitation could only be applied externally on the surface of the structures after crack formation occurs. This methodological necessity was mainly due to the limited lifetime (hours to a few days) of the (urease-based) enzymatic activity and/or viability of the applied bacterial species. In the present study, the application of alkali-resistant endospore-forming bacteria to enhance the self-healing capacity of concrete is investigated. Tensile and compressive strength characteristics of reference (no bacteria added) and bacterial concrete are quantified. Furthermore, the viability of bacteria immobilization in

Fig. 5.21 Biominerals observed by ESEM.

concrete is quantified and, finally, calcite precipitation potential of bacterial concrete is demonstrated by ESEM analysis (Figure 5.21).

Extensive results of this study are published elsewhere [50]. To date, the main conclusions of this ongoing research are that the alkaliphilic endospore-forming bacteria integrated in the concrete matrix can actively precipitate calcium carbonate minerals. Water needed for the activation of endospores can enter the concrete structure through freshly formed cracks. Furthermore, for mineral precipitation, active cells need an organic substrate that can metabolically be converted to inorganic carbon that can subsequently react with free calcium and form the precipitate, calcium carbonate. Free calcium is usually present in the concrete matrix, but the organic carbon is not. In the present experiments, organic carbon was externally applied as a part of the incubation medium, while ideally it should also be part of the concrete matrix. In this case, water needed is also applied externally to activate the concrete-immobilized bacteria, which can then convert organic carbon present in the concrete matrix to calcium carbonate and by doing so seal freshly formed cracks. We are currently investigating which specific kind of organic compounds are suitable to include in the concrete matrix. This is certainly not trivial as such compounds should be a suitable food source for bacteria as well as be compatible with concrete. Certain classes of organic compounds are less or not suitable at all, for example, compounds such as carbohydrate derivatives that are known to inhibit the setting of concrete even at low concentrations. Furthermore, we are presently investigating the long-term viability and potential possibilities to increase the viability of concrete-immobilized endospores to ensure long-lasting bacterially enhanced self-healing.

To conclude, we can state that the bacterial approach has potential to contribute to the self-healing capacity of concrete. We have shown that bacteria incorporated in high numbers (10^9 cm^{-3}) do not affect concrete strength, a substantial number of added bacteria remain viable, and moreover, these viable bacteria can precipitate calcium carbonate needed to seal or heal freshly formed cracks.

5.4
Self-healing Research at Cardiff

5.4.1
Introduction

The self-healing of cementitious materials has been the subject of a programme of study undertaken at Cardiff University in the United Kingdom since 2006. As a part of this programme the ability to artificially create a cementitious material with autonomic healing properties has been examined. The autonomic healing concept investigated in this work is illustrated in Figure 5.22. The concept is based on the principle that crack formation in the cementitious matrix also causes brittle adhesive-filled capsules or tubes, embedded within the matrix, to crack. The contents of the capsules or tubes are thereby released into the crack plane where the agent is intended to cure and heal the damaged host matrix. This concept was originally proposed for cementitious materials by Dry [10], as described in Section 5.2.1.3.

The long-term aim of this work is to establish a numerical material model for the autonomic healing process within cementitious materials. Given the limited availability of experimental data on this relatively new "smart" material, it has been deemed necessary to first undertake a series of experiments for which all data,

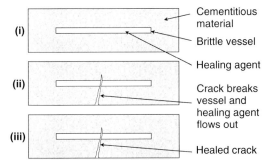

Fig. 5.22 Autonomic healing concept.

including (i) width, nature, and location of cracks; (ii) viscosity of adhesive; (iii) degree of migration of adhesive; and (iv) strength and stiffness of specimen pre- and post-healing, would be available.

This section, therefore, gives a brief overview of the early experimental work that has been conducted at Cardiff. This includes the preliminary investigation work that has led to the subsequent development of a successful autonomic healing experimental procedure. Initial results from this programme of study are presented and discussed, and some early conclusions and a summary of future work to be undertaken are given.

The work presented in this section is taken in part from results published in the first international conference on self-healing materials [51] and the coauthors doctorate thesis [52]. The reader is referred to the latter reference for further information on this topic.

5.4.2
Experimental Work

To date, the self-healing experimental research conducted at Cardiff has focused specifically on the development of adhesive-based healing within small-scale steel reinforced prismatic plain mortar beams. The work is similar in principal to that undertaken by Li et al. [25]; however, the concept has been extended for use on more traditional steel reinforced cementitious materials rather than fiber-reinforced ECCs.

5.4.2.1 Preliminary Investigations
Various aspects have been considered as part of the preliminary investigation stage, including the (i) type of host cementitious matrix, (ii) quantity of steel reinforcement, (iii) type of healing agent used, and (iv) method of agent encapsulation.

In order to investigate the feasibility of the self-healing concept, small-scale (255 × 75 × 75 mm) prismatic beams were used, as illustrated in Figure 5.24. In order to achieve a more homogeneous material at this specimen scale, beams were cast using mortar rather than concrete. The mix ratio adopted was 0.6:1:3.5 (water:OPC:sand). The sand aggregate used had a maximum particle size of 2 mm, which gave a minimum specimen dimension to maximum aggregate particle size ratio of 1:35.

In order to control crack development, but also allow sufficient crack opening for the adhesive healing agent to enter the crack plane before curing, the beams were reinforced with a single 3.15-mm-diameter high yield steel bar. This provides a cross-sectional area of steel close to the minimum EC2 code requirement of 0.15%.

The main types of healing agent which have been proposed to date in the literature, as discussed in Section 5.2.3.1, are (i) epoxy resins, (ii) cyanoacrylates, and (iii) alkali–silica solutions. In addition to being readily available and cost effective, a suitable agent for the autonomic healing of cementitious materials should

1. have the ability to be readily encapsulated;
2. have sufficient mobility to reach crack locations following release;
3. have sufficient mechanical properties postcuring to resist crack reopening; and
4. have sufficient long-term durability and compatibility with the cementitious matrix.

One type of epoxy resin and two types of cyanoacrylate (superglue) have been evaluated during the preliminary investigation stage. The epoxy resin tested was a Tecroc products injection grout TG07 and the cyanoacrylates examined were SICOMET 9000, a methoxyethyl-based product from Sichel-Werke GmbH, and Rite Lok EC-5, an ethyl-based cyanoacrylate produced by 3M Ltd.

Low-viscosity epoxy resins, such as Tecroc TG07, are two-part compounds, which currently form the principle healing agent used in the postdamage "active" remediation of critical concrete members such as floors and bridge decks. This product was found to be unsuitable for internal encapsulation, since the curing of the resin, which is initiated by the mixing of the two agents, was observed to continue despite being sealed within a capillary tube. In addition, the resin has a viscosity of 200 cPs at room temperature, and is therefore only capable of filling cracks down to 100 μm [53]. Separate storage of the two compounds in adjacent vessels has been investigated previously by Mihashi et al. [7]. However, the authors determined that this method offered poor curing properties due to insufficient mixing of the compounds, and therefore this option has not been pursued in this investigation.

Cyanoacrylates are one-part adhesives that cure in the presence of moisture and are noted for their ability to set rapidly within a period of seconds and provide a bond strength that often exceeds the strength of the substrate, certainly in the case of concrete. Of the two cyanoacrylates tested, Rite Lok EC-5 was selected as offering the greatest potential for self-healing the cementitious matrix. This was primarily due to its very low viscosity (1–10 cPs compared to 15–25 cPs for SICOMET 9000), and its general suitability for bonding a wide range of materials, including ceramic. Rite Lok EC-5 has a tensile strength (ISO 6922) of 20 MPa and a full curing time of 24 h. Full technical details of this adhesive are given by 3M [54].

With respect to the method of agent encapsulation, the most appealing delivery system from a practicality point of view is the containment of adhesive within

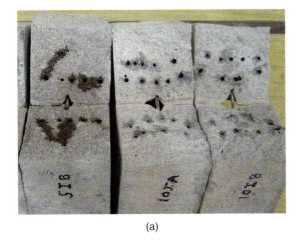

(a)

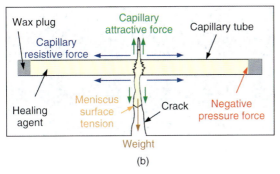

(b)

Fig. 5.23 (a) Ink staining on crack faces of control specimens from preliminary experiments and (b) schematic illustration of main forces acting on an internally encapsulated healing agent.

microcapsules that are embedded within the matrix during the casting stage, as employed for self-healing polymers by White *et al.* [55]. There are, however, significant production issues that must first be addressed before microcapsules can be used within cementitious composite materials. Therefore, in order to investigate the feasibility of the system, adhesive-filled hollow capillary tubes were used here in a manner similar to Li *et al.* [25]. Following a series of trial tests conducted on tubes of 0.8, 1.5, and 3 mm internal diameter, the latter was adopted due to its greater robustness during the casting process, and the lower capillary resistive force (Figure 5.23b) that it provides on the glue when it is in the process of flowing into the crack.

The preliminary experiments consisted of two sets of six beams, with each set comprising four self-healing (SH) and two control (C) beams tested under three-point loading. The first set contained a single layer of 5no. 100-mm-long capillary tubes, spaced evenly at a height of 20 mm from the bottom of the beam, and the second set contained two layers of 5no. tubes spaced evenly at heights of 20 and 35 mm from the bottom of the beam. The tubes in the SH beams were

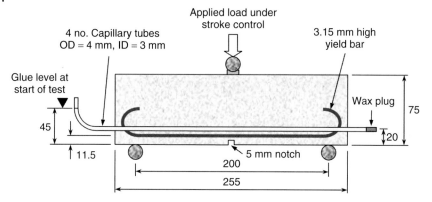

Fig. 5.24 Testing arrangement for self-healing experiments on notched beams *(Note: all dimensions are in mm and beam breadth = 75 mm).*

filled with Rite Lok EC-5 cyanoacrylate using a syringe and sealed with wax plugs, before casting. The tubes in the control beams were filled with an ink tracing die in a similar manner. The remainder of the preliminary testing arrangement, including the size and position of the steel reinforcement and notch, and the loading configuration were identical to the final experimental procedure described in the next section and illustrated in Figure 5.24.

There was evidence of a small amount of healing in one of the four SH beam tests with a double layer of tubes, but overall it was concluded that the glue had not been drawn into the cracks in sufficient quantity to allow for any significant amount of healing to occur in the SH beams. This conclusion was supported by the very limited extent of ink penetration, which was observed on the crack faces of the control beams, as illustrated in Figure 5.23a. It should also be noted that the glue has a slightly higher viscosity than the ink, and is therefore likely to have penetrated the crack to an even lower degree than the ink. No obvious glue deposition was observed on the crack faces of the SH beams, but large quantities of liquid glue and ink were observed to remain in both sections of the capillary tubes following complete fracture of all of the specimens.

The primary reason for this behavior is believed to be the inability of the capillary attractive force of the crack opening on the glue to overcome the large negative pressure forces created by the wax sealing plugs, as illustrated in Figure 5.23b. In order to eliminate these negative pressure forces, the final experimental procedure incorporated longer tubes projecting through the ends of the beams, which were open to the atmosphere, as shown in Figure 5.24.

5.4.2.2 Experimental Procedure

As a result of the preliminary investigation work described in the previous section, the following experimental procedure was established. This procedure was considered to be the most suitable method for gathering key data for the development of a model for adhesive-based self-healing within reinforced mortar prismatic beams.

Central deflection transducer

20 kN load cell

Support armature for transducer

4 no. adhesive supply tubes

CMOD clip gauge

Base support block

Fig. 5.25 Experimental arrangement.

A notched set of experiments consisting of four self healing (SH) and two control (C) beams were initially tested under three-point loading, as illustrated in Figure 5.24 and shown in Figure 5.25.

Four 300-mm-long capillary tubes were cast into each mortar beam at 15 mm centers in the transverse direction and at a height of 20 mm from the bottom of the beam. These tubes were inserted through lubricated predrilled holes in the mould end plates, and following casting, curved plastic supply tubes were attached to one end of each tube, as shown in Figure 5.25. All beams were reinforced with a single 3.15-mm-diameter high yield steel rod placed on 10 mm spacer blocks and anchored within the mortar by hooked ends. All specimens were demoulded 24 h after casting and air cured for a period of 28 days before testing.

In preparation for testing, the four capillary supply tubes in each of the SH beams were filled with Rite Lok EC-5 to a level of 25 mm above the centre-line of the straight tube before being plugged at one end only with a wax seal. The two control beams were filled and sealed in a similar fashion, but with an ink tracing die of low viscosity ($\sim$3 cPs).

A series of three-point bend tests were then completed in a Shimadzu AG-1 test rig fitted with a 20 kN load cell, as illustrated in Figure 5.25. All tests were controlled by machine stroke, at a speed of 0.2 mm min^{-1}. The load, crack mouth opening displacement (CMOD), and central deflection were recorded on a multichannel Orion data logger.

The testing procedure involved the four SH beams being loaded beyond the point of first pronounced cracking and then up to a CMOD of approximately 0.55 mm. The load was then removed and the beam was left to fully heal for 24 h prior to being remounted in the test rig and tested to failure. The two control specimens, with ink filled tubes, were tested in a similar manner and at the same loading rate, but in a single test, since it was unnecessary to allow for any healing period.

5.4.3
Results and Discussion

Three of the four SH beams tested in this series of experiments showed significant evidence of a self-healing behavior, and it should be noted that the limited healing

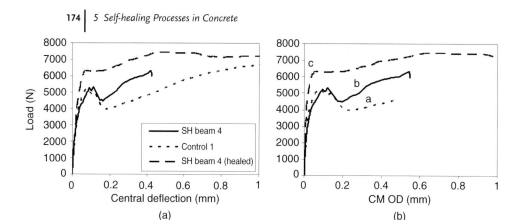

Fig. 5.26 (a) Load-central deflection and (b) load-CMOD response for SH beam 4.

exhibited by the remaining beam can be directly attributed to problems associated with the delivery system for that particular beam.

The load-central deflection and load-CMOD response of a representative SH beam (SH 4) and a control beam are presented in Figure 5.26. It may be seen that for both the control specimen and the first test on SH beam 4, there is some prepeak nonlinearity due to microcracking between about 4 and 5.2 kN. This is followed by a sudden drop of approximately 0.7 kN over a CMOD increase of approximately 0.1 mm for the SH beam compared with about 1.1 kN over 0.13 mm for the control beam. This drop is caused by the brittle fracture of the 4 no. borosilicate capillary tubes which emit a distinctive breaking sound during testing. Thereafter, the primary load carrying mechanism is performed by the steel reinforcement and the stiffness of the control beam is given by the gradient of the line at point "a" in Figure 5.26b. It may be observed, however, that even during the first test on SH beam 4 there is some evidence of "primary" healing, reflected by the increased gradient of the line at point "b". Cyanoacrylates are acidic solutions that have the ability to cure rapidly within a period of seconds. It is believed that the conditions within the mortar, including the presence of moisture and the alkaline environment, further accelerate the curing process, and are responsible for this rapid primary healing.

There is also clear evidence of the occurrence of a "secondary" healing effect if the mechanical response of SH beam 4, tested after a period of 24 h, is compared to that of the original control specimen; the gradient of the response up to point "c" is stiffer than that of the original control beam, the peak has increased by over 20%, and the postpeak response is more ductile.

Therefore, it would appear that the infiltration of glue into the initial crack has a self-enhancing rather than self-healing effect. It is believed that the cyanoacrylate is not only adhering the original crack faces together but is also infiltrating the microcracked zone of damage around these crack faces owing to its very low viscosity, thereby creating a localized zone of a polymer–cementitious composite material. This composite material has enhanced mechanical properties when compared to the plain cementitious material. Qualitative evidence of the healing

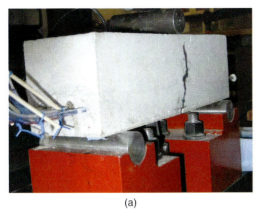

(a)   (b)

Fig. 5.27 (a) Ink staining during control test and (b) glue seepage during SH test.

effect of the cyanoacrylate on mortar is shown in Figure 5.28a, whereby the final through crack on the side surface of SH beam 3 can be seen to propagate along a different path when compared to the original crack caused during the first stage of loading on this beam.

During the experiments, evidence of the movement of ink from the capillary tubes into the crack was observed. Shortly after the initial fracture of the tubes within the control specimens, staining patterns on the side of the beams became apparent, as shown in Figure 5.27a. Glue was also observed to flow out of the crack during the first stage of testing for three of the four SH beams, and also during the second stage of testing after 24 h (Figure 5.27b). This illustrates the fact that when stored in sufficient volume under ordinary room conditions cyanoacrylate does not set quickly despite being exposed to the atmosphere. This introduces the possibility of achieving further healing during subsequent tests, although this has not been examined as part of this initial investigation.

Examination of the crack faces after testing indicated the presence of glue above and below the level of the capillary tubes, as illustrated in Figure 5.28b. This suggests that despite the fast speed with which the cyanoacrylate is believed to cure in the mortar, it remains in a fluid state for a period sufficient enough to rise under the influence of capillary action within the crack, and also flow downward under both capillary action and gravity. It should also be noted that, following specimen fracture, large amounts of both glue and ink were observed to remain at the ends of the tubes that had been sealed, thus confirming the findings of the preliminary investigations in respect to the restrictive effect of the negative pressure force on the flow of glue caused by the wax seals.

5.4.4
Modeling the Self-healing Process

The self-healing mechanism occurring during the experimental investigation described above is a complex combination of physical and chemical processes.

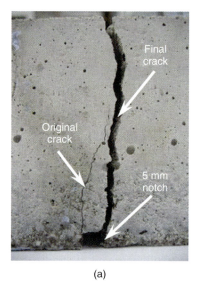

(a)

(b)

Fig. 5.28 (a) New crack formation on side face of SH beam 3 and (b) effective zone of healing indicated by glue migration on crack faces of SH beam 4.

After the cracking of the mortar and the breakage of the glass tubes, the adhesive is pulled into the crack under the influence of capillary attractive forces created by the crack opening. In reality, however, this capillary force gradually diminishes since the width of the propagating vertical crack at the location of the tubes is constantly increasing. In addition, the adhesive becomes more viscous as it begins to cure. The curing process in itself is dependent on many factors, including ambient temperature, moisture content and alkalinity of the mortar, and thickness of the glue layer.

This complex physical–chemical process may be captured through the development of a complex nonlinear coupled phenomenological-based constitutive model. Alternatively, the basic mechanical response of the healing process may be captured in a far simpler fashion using a discrete modeling method. Recent work has therefore focused on the use of the lattice beam modeling method [56], originally used for the fracture modeling of cementitious materials by Schlangen and van Mier [57].

The lattice modeling method is based on the discretization of a continuum into a network of one-dimensional beam elements. Beam properties are assigned either according to the respective material phases that the beams represent (e.g. mortar, aggregate, and interfacial zone) or according to a defined statistical distribution. Cracks are modeled discretely in this approach and thus crack openings are determined automatically. This method has been shown to capture the softening branch of fracture tests on quasi-brittle materials, despite assuming a simple elastic–brittle constitutive relationship for the beam elements [58].

A crack width criterion is used to govern the breakage of the capillary tubes within the model. Once the capillary tubes have broken, it is assumed that the flow of the adhesive is controlled by the varying aperture of the crack at that location. The approach utilized within this model is therefore based on capillary flow through nonuniform sections [59] and is similar to that employed by Roels *et al.* [60] for modeling moisture flow in discrete cracks in building materials. This involves using a 1D moving front model in which both capillary and gravity forces are considered. The governing equations are Darcy's equation of flow and mass continuity, with a sink term included in the latter to model the flow of glue into the fracture process zone surrounding the macrocrack.

Glue setting is modeled in a staged manner whereby broken beams are replaced with "healed" beams at predefined CMOD limits. The height of glue rise, and therefore determination of which broken beams are to be healed, is obtained from the nonuniform capillary flow theory described above. The healed beams are considered to be composite beams comprising part mortar and part glue. The axial stiffness of these composite beams is therefore determined from the axial stiffness of the mortar part and the glue part combined in series. The length of the glue part is determined by the width of the opening at the beam location prior to healing.

Initial results indicate that the lattice beam modeling method is capable of being easily modified to allow for localized strength and stiffness increase within the specimens due to healing. This is achieved through reintroducing previously broken beams back into the simulation with newly defined properties. The model is therefore capable of capturing not only the decrease in specimen stiffness due to microcracking of the mortar in the initial stages of a test but also the increase in stiffness due to primary healing, as illustrated in Figure 5.26. For further information on this modeling approach see Joseph [52].

5.4.5
Conclusions and Future Work

The findings of an initial investigation examining adhesive-based self-healing within lightly reinforced prismatic mortar beams, subject to three-point bending, have been presented in this section. The qualitative and quantitative results have indicated that an ethyl cyanoacrylate-based healing agent supplied via an external supply system is capable of achieving a successful self-healing mechanism. These results have, therefore, confirmed the feasibility of adhesive-based healing of cementitious materials as originally investigated by Dry [10].

The postcracked stiffness, peak load, and ductility were all observed to increase posthealing when the mechanical responses of the healed beams were compared to those of the control specimens. In addition, a primary healing effect was also observed to occur during the first stage loading of the beams containing adhesive. This was identified by an increased stiffness response of the beam following the distinct cracking of the capillary tubes.

Observations made during and after testing clearly indicate that, provided the healing agent is open to the atmosphere and has sufficiently low viscosity, ethyl cyanoacrylate glue is capable of penetrating a significant area of the crack surface under the influence of capillary suction forces and gravity. Successful bonding of the original crack faces has also been identified by the formation of new crack paths during testing of the specimens posthealing.

Preliminary modeling work has also indicated that the discrete lattice beam modeling method, modified to allow for beam healing, is capable of capturing the basic mechanical response of the self-healing system.

It is clear that the main disadvantage of the current adhesive-based self-healing procedure outlined in this section is the issue of practicality. Extending the current laboratory-based methodology to full size structural members cast on-site is not currently feasible. However, future development of a more sophisticated delivery network that has sufficient flexibility to survive the initial casting process, but which becomes brittle postencasement would greatly enhance the feasibility of this self-healing technique.

As part of the experimental future work programme, the effect of loading rate on the amount of primary healing observed during the first test for glue filled mortar beams with different levels of steel reinforcement is to be investigated. In addition, glue uptake values for different crack opening widths are to be quantified, and the extent of adhesive migration into the microcracked zone around the crack face is to be established. With respect to numerical modeling, further development of the lattice model to allow for the nonlinear effects of reinforcement slip and yielding are to be implemented.

5.5
A View to the Future

Concrete structures are always cracked to some degree. These cracks, especially when they amalgamate, provide pathways to moisture and fluid ingress that subsequently contribute to the erosion of the concrete through various chemical and physical processes. The durability issues associated with concrete are significant and are not going to reduce in the near future, unless materials with rejuvenating healing abilities are researched, developed, and promoted. Since the manufacture of 1 t of PC produces an approximate equivalent weight of CO_2 [1], and given the unprecedented level of public consciousness regarding the problems of global warming, there has never been a more opportune time for the development of these materials. For this to occur, clients must begin to take responsibility for evaluating long-term whole lifetime costs, rather than just focusing on the short-term construction costs when they commission their structures. These costs should be evaluated in both financial and environmental terms. Government legislation will probably be required to influence such a seed change in thinking. However, it is the view of the authors that this is a priority area for Governments, because it has the potential to begin to break the link between economic growth and environmental degradation.

5.6
Acknowledgments

The first author would like to acknowledge his colleagues in the Microlab, that is, Klaas van Breugel, Henk Jonkers, Mario de Rooij, Ye Guang, Eddy Koenders, Oguzhan Copuroglu, and Nienke Ter Heide, for their inspiring discussions on the subject of self-healing and for allowing publishing part of their work. Furthermore, discussions with Victor C. Li and Sybrand van der Zwaag are highly appreciated.

The second author would like to acknowledge Bhushan Karihaloo for originally introducing him to the concept of autonomic healing and Tony Jefferson, Mauro Cantoni, Ben Isaacs, and Cardiff School of Engineering laboratory personnel for their assistance in the development and undertaking of the self-healing work conducted at Cardiff University. The author is also grateful for the support and supply of cyanoacrylate from 3M Ltd.

References

1 van Oss, H.G. (2005) "Background Facts and Issues Concerning Cement and Cement Data". Open-file report 2005-1152, U.S. Department of the Interior and U.S. Geological Survey.

2 van Breugel, K. (2007) "Is There a Market for Self-healing Cement-based Materials?" 1st International Conference on Self-healing Materials, Noordwijk, Holland.

3 Yunovich, M. and Thompson, N.G. (2003) *Concrete International*, 25(1), 52–57.

4 Freyermuth, C.L. (2001) *Concrete International*, 23(2), 89–95.

5 DTI (2006) *Construction Statistics Annual Report 2006*, TSO, London.

6 DEFRA. (2008) *Joint UK-Sweden Initiative on Sustainable Construction*. Available at: <http://www.constructing excellence.org.uk/uksweden/defra.jsp? level = 0>* [Accessed: 17th April, 2008].

7 Mihashi, H., Kaneko, Y., Nishiwaki, T. and Otsuka, K. (2000) *Transactions of the Japan Concrete Institute*, 22, 441–50.

8 Sharp, S.R. and Clemeña, G.G. (2004) *State of the Art Survey of Advanced Materials and Their Potential Application in Highway Infrastructure*, Virginia Transportation Research Council, Charlottesville, pp. 1–41.

9 Schmets, A.J.M. (2003) Self-healing: An Emerging Property for New Materials, [Web Page], Leonardo Times, Available at: <*http://www.tudelft.nl/ live/binaries/ca7822ae-5305-4162-8477-28a88507da1f/doc/Article% 20Self%20Healing%20Materials.pdf*> [Accessed: 23rd April 2008].

10 Dry, C. (1994) *Smart Materials and Structures*, 3(2), 118–23.

11 Westerbeek, T. (2005) Self-healing Materials Radio Netherlands. Available at: <*http://www2.rnw.nl/rnw/en/ features/ science/050801rf?view = Standard*> [Accessed: 25th November, 2005].

12 Schlangen, E. (2005) Self-healing Phenomena in Cement-based Materials, [Web Page]. RILEM. Available at: <*http://www.rilem.org/tc_shc.php*> [Accessed: 29th November 2005].

13 Kishi, T., Ahn, T., Hosoda, A., Suzuki, S. and Takaoka, H. (2007) "Self-Healing Behaviour by Cementitious Recrystallisation of Cracked Concrete Incorporating Expansive Agent", 1st International Conference on Self-healing Materials, Noordwijk, Holland.

14 Zhong, W. and Yao, W. (2008) "Influence of damage degree on self-healing of concrete". *Construction and Building Materials*, 22(6), 1137–1142.

15 Reinhardt, H.-W. and Jooss, M. (2003) "Permeability and self-healing of cracked concrete as a function of temperature and crack width". *Cement and Concrete Research*, 33, 981–85.

16 Şahmaran, M. and Li, V.C. (2008) *Cement and Concrete Composites*, 30, 72–81.

17 Jacobsen, S. and Sellevold, E.J. (1996) *Cement and Concrete Research*, 26(1), 55–62.

18 Edvardsen, C. (1999) *ACI Materials Journal*, 96(4), 448–54.

19 Ter Heide, N., Schlangen, E. and van Breugel, K. (2005) "Experimental study of crack healing of early age cracks", in (eds. Jensen, O.M., Geiker, M. and Stang, H.), 311–22 *Proceedings*

Knud Højgaard Conference on Advanced Cement-Based Materials, Technical University of Denmark.

20 Hosoda, A., Kishi, T., Arita, H. and Takakuwa, Y. (2007) Self Healing of Crack and Water Permeability of Expansive Concrete. "1st International Conference on Self-healing Materials", Noordwijk, Holland.

21 Yamada, K., Hosoda, A., Kishi, T. and Nozawa, S. (2007) *Crack Self Healing Properties of Expansive Concretes with Various Cements and Admixtures*. 1st International Conference on Self-healing Materials, Noordwijk, Holland.

22 Dry, C.M. (2000) *Cement and Concrete Research*, 30(12), 1969–77.

23 Dry, C. and Corsaw, M. (2003) *Cement and Concrete Research*, 33(11), 1723–27.

24 Dry, C. and Unzicker, J. (eds) (1998) *Smart Structures and Materials 1998 Smart Systems for Bridges, Structures, and Highways*, The International Society for Optical Engineering, San Diego.

25 Li, V.C., Lim, Y.M. and Chan, Y.W. (1998) *Composites Part B: Engineering*, 29(B), 819–27.

26 Nishiwaki, T., Mihashi, H., Jang, B.-K. and Miura, K. (2006) *Journal of Advanced Concrete Technology*, 4(2), 267–75.

27 Springenschmidt, R. (1994) "Thermal cracking in concrete at early-ages", *Proceedings of the International Rilem Symposium no. 25*, E and FN Spon, London.

28 Salet, T. and Schlangen, E. (1997) "Early-age crack control in tunnels", in *Proceedings of Euromat 97*, Vol. 4, (eds Sarton, L. and Zeedijk, H.), Maastricht, 367–77.

29 *www.dcmat.tudelft.nl* [Accessed: 2008].

30 Jonkers, H.M. and Schlangen, E. (2007) "Self-healing of cracked concrete: a bacterial approach", in (eds Carpenteri, *et al.*) *Proceedings of FRAMCOS6: Fracture Mechanics of Concrete and Concrete Structures*, Catania, Italy, 17–22 June 2007; p: 1821–26.

31 www.senternovem.nl/IOP_ Selfhealingmaterials [Accessed: 2008].

32 Ter Heide, N. (2005) Crack Healing in Hydrating Concrete, MSc-thesis, Delft University of Technology, The Netherlands.

33 Granger, S., Loukili, A., Pijaudier-Cabot, G. and Chanvillard, G. (2005) Mechanical characterization of the self-healing effect of cracks in Ultra High Performance Concrete (UHPC), in Proceedings Third International Conference on Construction Materials, Performance, Innovations and Structural Implications, ConMat'05, Vancouver, Canada, August 22–24.

34 www.femmasse.com [Accessed: 2008].

35 Wang, K., Jansen, D.C., Shah, S.P. and Karr, A.F. (1996) "Permeability Study of Cracked Concrete". Technical Report 46, National Institute of Statistical Sciences.

36 Jorgensen, B.B. and D'Hondt, S. (2006) Science, 314, 932–34.

37 Dorn, R.I. and Oberlander, T.M. (1981) Science, 213, 1245–47.

38 De la Torre, J.R., Goebel, B.M., Friedmann, E.I. and Pace, N.R. (2003) Environmental Microbiology, 69(7), 3858–67.

39 Fajardo-Cavazos, P. and Nicholson, W. (2006) Applied and Environmental Microbiology, 72(4), 2856–63.

40 Pedersen, K., Nilsson, E., Arlinger, J., Hallbeck, L. and O'Neill, A. (2004) Ex-tremophiles, 8(2), 151–64.

41 Sleep, N.H., Meibom, A., Fridriksson, T., Coleman, R.G. and Bird, D.K. (2004) Proceedings of the National Academy of Sciences of the United States of America, 101(35), 12818–23.

42 Sagripanti, J.L. and Bonifacino, A. (1996) Applied and Environmental Microbiology, 62(2), 545–51.

43 Schlegel, H.G. (1993) General Microbiology, 7th edn, Cam-bridge University Press.

44 De Graef, B., De Windt, W., Dick, J., Verstraete, W. and De Belie, N. (2005) Materials and Structures, 38(284), 875–82.

45 Ghosh, P., Mandal, S., Chattopadhyay, B.D. and Pal, S. (2005) Cement and Concrete Research, 35(10), 1980–83.

46 Dick, J., De Windt, W., De Graef, B., Saveyn, H., Van der Me-eren, P., De Belie, N. and Verstraete, W. (2006) Biodegradation, 17(4), 357–67.

47 Rodriguez-Navarro, C., Rodriguez-Gallego, M., Ben Chekroun, K. and Gonzalez-Munoz, M.T. (2003) Applied and Environmental Microbiology, 69(4), 2182–93.

48 Bang, S.S., Galinat, J.K. and Ramakrishnan, V. (2001) Enzyme and Microbial Technology, 28, 404–9.

49 Ramachandran, S.K., Ramakrishnan, V. and Bang, S.S. (2001) ACI Materials Journal, 98(1), 3–9.

50 Jonkers, H.M. (2007) Self healing concrete: a biological approach, in Self Healing Materials (ed. S. vanderZwaag), Springer.

51 Joseph, C. and Jefferson, A.D. (2007) Issues Relating to the Autonomic Healing of Cementitious Materials, 1st International Conference on Self-healing Materials, Noordwijk, Holland.

52 Joseph, C. (2008) Experimental and Numerical Study of Fracture and Self Healing of Cementitious Materials, PhD thesis, Cardiff University.

53 Tecroc Ltd. (2004) Technical data sheet 1.3.1—Epoxy Injection Grout, Issue date: Dec 2004. Available at: http://www.tecroc.co.uk/Technical_ Data_Sheets/15_EPOXY_ INJECTION_GROUT.pdf.

54 3M Ltd. (2007) Product data sheet—Rite-Lok Cyanoacrylate Adhesive EC5, Issue date: Feb 2007.

55 White, S.R., Sottos, N.R., Geubelle, P.H., Moore, J.S., Kessler, M.R., Sriram, S.R., Brown, E.N. and Viswanathan, S. (2001) Nature, 409, 794–97.

56 Joseph, C., Jefferson, A.D. and Lark, R.J. (2008) Lattice Modelling of Autonomic Healing Processes in Cementitious Materials, WCCM8/ECCOMAS 2008, Venice, Italy.

57 Schlangen, E. and van Mier, J.G.M. (**1992**) *Cement and Concrete Composites*, **14**(2), 105–18.

58 Joseph, C. and Jefferson, A.D. (**2007**) *Stochastic Regularisation of Lattice Modelling for the Failure of Quasi-brittle Materials, Proceedings of FRAMCOS6: Fracture Mechanics of Concrete and Concrete Structures*, Catania, July, 17–22 June 2007; p: 445–52.

59 Young, W.-B. (**2004**) *Colloids and Surfaces A. Physicochemical and Engineering Aspects*, **234**, 123–28.

60 Roels, S., Vandersteen, K. and Carmeliet, J. (**2003**) *Advances in Water Resources*, **26**, 237–46.

6

Self-healing of Surface Cracks in Structural Ceramics

Wataru Nakao, Koji Takahashi and Kotoji Ando

6.1
Introduction

Self-healing is the most valuable phenomenon to overcome the integrity decreases that are caused by the damages in service. Thus, self-healing should occur automatically as soon as the damages occur, and the healed zone should have high integrity as it was before damaging. When proposing self-healing materials, one must know the nature of the damage and the service conditions of the materials. In the case of the structural ceramics, the most sever damage is surface cracks, which is possible to be introduced by crash, fatigue, thermal shock, and corrosion during their service time. Over the past 30 years, ceramics have become the key materials for structural use at high temperature due to their enhanced quality and good processability. Structural ceramics are also expected to be applied in the corrosion environments such as air, because of its chemical stability. Thus, self-healing of surface cracks in the structural ceramics is an important issue to ensure the structural integrity of ceramic components.

In this chapter the mechanism and effects of self-healing of surface cracks in structural ceramics are introduced. Apart from this, the fracture manner of ceramics is also discussed. This will help the readers to understand the self-healing phenomena in ceramics and its benefits. The history of crack healing is also included in the text. Furthermore, new methodology to ensure the structural integrity using crack-healing effect and advanced ceramics having self–crack-healing ability are mentioned.

6.2
Fracture Manner of Ceramics

Ceramics tend to have brittle fracture that usually occurs in a rapid and catastrophic manner. Brittle fracture is usually caused by the stress concentration at the tip of the flaws. For brittle fracture under pure mode, I, loading, under which crack is

Self-healing Materials: Fundamentals, Design Strategies, and Applications. Edited by Swapan Kumar Ghosh
Copyright © 2009 WILEY-VCH Verlag GmbH & Co. KGaA, Weinheim
ISBN: 978-3-527-31829-2

subjected to opening, the failure criterion is that the stress intensity factor, K_I, is equal to the fracture toughness, K_{IC}. The stress intensity factor is an indicator of the magnitude of stress near a crack tip or the amplitude of the elastic field. The value of K_I can be obtained from the liner elastic mechanics as follows:

$$K_I = \sigma \cdot Y \sqrt{\pi \cdot a} \tag{6.1}$$

where a is the crack length, σ is the tensile stress applying to crack perpendicularly, and Y is a dimensionless parameter that is determined from the crack and loading geometries. Thus the stress at failure is given by

$$\sigma_c = \frac{1}{Y} \cdot \frac{K_{IC}}{\sqrt{\pi \cdot a}} \tag{6.2}$$

From the failure criterion, one can understand that the fracture strength of ceramic components is not an intrinsic strength but is fracture toughness and crack geometry.

In general, ceramic components contain flaws at which the stress concentration causes brittle fracture before their end-use. These flaws have particular figurations and sizes and are introduced mainly during manufacturing. Figure 6.1 shows a schematic diagram of the flaw populations that could exist in ceramics. In this example, the most severe flaws are surface cracks, perhaps resulting from machining. The next most severe flaws are voids and pores introduced by sintering. Voids and pores would become the main fracture of origin, if the surface cracks are much smaller or removed, such as healed. The variation in the failure sources leads to the large strength distribution in ceramics.

There are few scenarios that can generate or introduce surface cracks during service. One possibility is by contact events, for example, impact, erosion, corrosion, and wear. Contact events may cause high stresses to the vicinity of the contact site, leading to crack formation. Sudden changes in temperature can also lead to stresses, known as *thermal stress* or *thermal shock*. The introduced surface cracks would be more severe than the pores and voids. As a result, these crackings cause large strength decrease to the component. If these scenarios occur again, it is possible that the component fractures catastrophically.

As mentioned above, ceramics usually fracture when $K_I = K_{IC}$ but, in studies of fracture [1–3], it is sometimes found that crack growth can occur at lower value

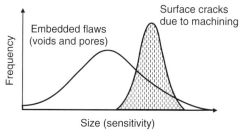

Fig. 6.1 Population of the strength-controlling flaws.

of K_I. The mechanisms have been analyzed to describe the slow crack growth behavior, including chemical reaction kinetics and interfacial diffusion. At the stressed crack tip, it is found that environmental species, for example, moisture, react and break the bonds at the crack tip resulting in stress corrosion cracking. The kinetics depends on the concentration of the environmental species. Actually it was found that the crack growth velocity in toluene is less than that in air [2]. This implies that the presence of moisture enhances the behavior. Furthermore, at high temperature, it is also found that the localized creep damage can give rise to the slow crack growth.

6.3
History

In 1966, the study on the strengthening behavior of ceramics by heat treatment was reported by Heuer and Roberts [4]. Then, Lange and Gupta [5] reported the strengthening of ZnO and MgO by heat treatment, and used the term "crack healing" for the first time in 1970. Now, we can find more than 200 reports on the strengthening effects by heat treatment for cracking ceramics. The crack-healing mechanisms in these reports can be roughly categorized into

1. re-sintering
2. relaxation of tensile residual stress at the indentation site
3. cracks bonding by oxidation.

Re-sintering [5–11], that is, diffusive crack-healing process, is an older crack-healing concept and commences with a degradation of the primary crack. This regression generates regular arrays of cylindrical voids in the immediate crack tip vicinity. Also, some studies [9–11] on the model and the kinetics of diffusive crack healing in single crystalline and polycrystalline ceramics have been proposed. However, as this crack healing requires the high crack-healing temperature, grain growth might also be generated. In some cases, the strength decreases than it before heat-treatment, although large strength recovery due to crack healing is attained. The relaxation of tensile residual stress at the indentation site leads to strength recovery. However, this phenomenon does not heal cracks. The crack bonding by oxidation has been first reported by Lange [12]. He investigated the strength recovery of the cracked polycrystalline silicon carbide (SiC) by heat treatment in air at 1673 K, and reported that the average bending strength of the specimens heat-treated for 110 h was 10% higher than that of the unheat-treated specimens. The same phenomenon in polycrystalline silicon nitride (Si_3N_4) was reported by Easler et al. [13]. The heat treatment temperature required for this crack-healing mechanism by oxidation is less than that required for the re-sintering crack healing. The other important aspect is that the cracks healed by this mechanism are filled with the formed oxides. A further mechanism and method for crack healing has been proposed. As an example, Chu et al. [14] proposed the crack-healing method using penetrating glasses. They succeeded to repair cracks in alumina (Al_2O_3) and have found that the repaired part becomes even stronger than the base alumina.

As mentioned above, many other investigators have shown their interest on crack healing of ceramics. In the field of ceramic nanocomposites, there are many reports also available on crack healing. The original impulsion for research in crack healing of ceramic nanocomposites originates from the works of Niihara and coworkers [15–17]. They observed that the strength of the alumina containing 5 vol% of submicrometer-sized SiC particles can be enhanced by annealing at 1573 K for 2 h in Argon. Since the original report, various mechanisms have been proposed to explain this phenomenon. Nowadays, this mechanism is confirmed to be driven by the oxidation of SiC particles. Thompson et al. [18] observed that partial healing of indentation cracks occurred when 5 vol% 0.15 μm SiC particles reinforced alumina were annealed at 1573 K for 2 h. Chou et al. [19] have also investigated the crack length and bending strength of alumina/5 vol% 0.2 μm SiC particles nanocomposite after annealing at 1573 K for 2 h in Argon or air, concluding that crack healing occurs by the oxidation of SiC particles. The similar conclusion was also derived by Wu et al. [20]. However, Chou et al. [19] noted that a uniform reaction layer was not formed between the crack walls, because the lower SiC content (only 5 vol%) results in a small quantity of the formed oxide.

Ando and coworkers observed that the similar crack healing in mullite ($3Al_2O_3$ $2SiO_2$) [21–23], Si_3N_4 [24–27] and alumina [28–31] based composites containing more than 15 vol% SiC particles can recover the cracked strength completely. They found that the healed zone is mechanically stronger than the base material and proposed the following requirements to obtain a strong healed zone:

1. Mechanically strong products (compared to the base material) should be formed by the crack-healing reaction.
2. The volume between crack walls should be completely filled with the products formed by the crack-healing reaction.
3. The bond between the product and crack wall should be strong enough.

Crack-healing reports can be classified into three generations, as shown in Table 6.1.

Table 6.1 Categorization of self-healing ceramics.

Types	Mechanism	Triggered by damage	Valid under service condition	Strong healed part
First generation	Re-sintering	No	No	Yes
Second generation	Oxidation of SiC (<5 vol% SiC)	Yes	Yes	No
Third generation	Oxidation of SiC (>10 vol% SiC)	Yes	Yes	Yes

First generation, that is, crack healing driven by re-sintering is only to recover the cracked strength. Second generation, that is, crack healing driven by oxidation of less than 10 vol% SiC can be triggered by damage and occur under service conditions, but the strength recovery is inadequate. Third generation, that is, crack healing proposed by Ando *et al.* can be attained with all the requirements. Consequently, the third generation crack healing is confirmed to be "true" self-healing for structural ceramics.

6.4
Mechanism

To keep the structural integrity of ceramics, an efficient self-healing should occur. This is possible if healing the surface cracks obeys the following conditions:
1. Healing must be triggered by cracking;
2. Healing must occur at high temperature [as structural ceramics are expected to typically operate at high temperature (~1273 K) in air] in the corrosion atmosphere, such as air;
3. Strength of the healed zone must be superior to the base material.

Self-crack healing driven by the oxidation of silicon carbide (SiC) can be qualitatively understood to satisfy requirements 1 and 2. Figure 6.2 shows the schematic of the crack healing in the ceramics containing SiC particles heated at high temperature in the presence of air. Cracking allowes SiC particles located on the crack

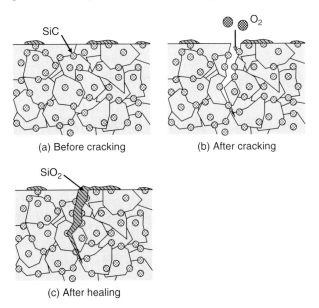

(a) Before cracking   (b) After cracking

(c) After healing

Fig. 6.2 Schematic illustration of crack-healing mechanism.

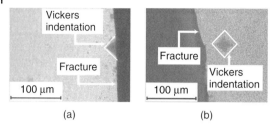

(a)   (b)

Fig. 6.3 Fracture initiation of alumina/15 vol% 0.27 μm SiC particles composite: (a) as-cracked and (b) crack healed at 1573 K for 1 h in the presence of air.

walls to react with the oxygen in the atmosphere resulting in healing. The details of the valid conditions are discussed later (Section 6.6). Subsequently, the crack is completely healed as oxidation progresses. As mentioned earlier, if the three important conditions of achieving strong healed zone is satisfied, then fracture initiation changes from the surface crack to the other flaws such as embedded flaw. This behavior is well demonstrated in Figures 6.3a and b [29].

The following equation showing oxidation of SiC also supports the above findings.

$$SiC + 3/2\, O_2 = SiO_2 + CO \tag{6.3}$$

There exist two important features in the above mentioned process. One is the increase in the volume of the condensed phase and the other is the generation of the huge exothermic heat. Because mole number of silicon is held constant during the oxidation, the volume increase is found to be 80.1%. As oxidation progresses, the crack walls are covered with the formed oxide. Finally, the space between the crack walls is completely filled with the formed oxide. For the complete infilling of the space between crack walls, it is necessary to contain more than 10 vol% SiC. (Section 6.5) Another important parameter for attaining the complete infilling is the size of crack.

From Figure 6.4 [31], one can find the critical crack size that can be completely crack healed. As an example, the critical crack size of alumina/30 vol% SiC particles composite is 300 μm. This value is the surface length of a semi-elliptical crack with an aspect ratio (crack depth/half of surface length) of 0.9 introduced by indentation method. Below this value, the crack-healed specimens exhibit the same strength, because the space between crack walls is completely filled with the formed oxide and because the fracture initiates from the embedded flaws. When the value is above the critical crack size, the space between the cracks walls is too large to be filled with the formed oxide.

Alternatively, the heat generated during the reaction makes the formed oxide and the base material to react or to once melt. The second low enthalpy change of the reaction can be evaluated to be −945 kJ using the thermochemical data [32] of the pure substance. This phenomenon might lead to strong bonding between

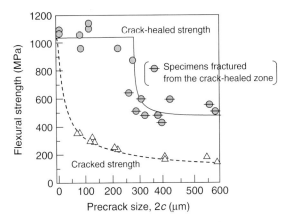

Fig. 6.4 Flexural strength of the crack-healed alumina/30 vol% 0.27 μm SiC particles composite as a function of surface length of a semi-elliptical crack.

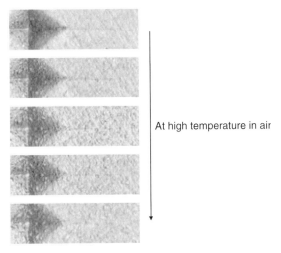

At high temperature in air

Fig. 6.5 Photographs of *in situ* observation of crack healing, in which alumina/15 vol% 0.27 μm SiC particles composite containing cracks with an indentation is heat-treated at 1573 K in the presence of air.

the reaction products and crack walls. The crack-healing mechanism is clearly demonstrated by *in situ* observation, as shown in Figure 6.5.

The phenomenon was observed while alumina/15 vol% 0.27 μm SiC particles composite containing indentation cracks is heat-treated at 1573 K in the presence of air. The features of the crack healing behavior are as follows: (i) the reaction products like sweats appear from the cracks and surface as the reaction progresses; (ii) cracks are perfectly covered and filled with the reaction products; (iii) the reaction products form with bubbling; and (iv) there are no changes in indentation

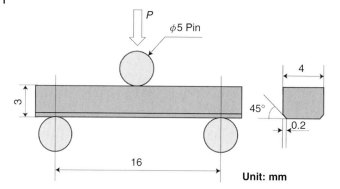

Fig. 6.6 Dimensions of three-point bending and the bar specimens.

figuration. From the observation, it is noted that the high temperature makes the reaction products, as well as base material melting, and the bubble, including carbon monoxide (CO) gas, forming strong crack-healed zone.

Furthermore, to estimate the strength of the healed zone in detail, it is necessary to take account of the following issues, that is:

(i) Effective volume should be so small that most fracture initiates at the crack-healed zone.

(ii) The strain energy at failure should be so low that fracture initiation is identified easily.

Ando and coworkers have adapted a three-point bending method as shown in Figure 6.6 for fracture tests.

The span of the geometry is 14 mm less than that of Japan Industry Standard (JIS) [33]. The crack-healed specimens have higher strength than the smooth mirror-polished specimens (Section 6.6). Polishing was carried out according to JIS standard [33].

6.5
Composition and Structure

6.5.1
Composition

Most important factor to decide the self–crack-healing ability is the volume fraction of SiC. As mentioned in Section 6.4, it is necessary for achieving strong crack-healed zone that the volume between crack walls is completely filled with the products formed by the crack-healing reaction. Therefore, there is lower limit of SiC volume fraction to endow with adequate self–crack-healing ability. Figure 6.7 shows the cracked and crack-healed strengths of alumina containing various volume fractions of SiC particles, which has mean particle size of 0.27 μm. As a result, the crack-healed strength varies with SiC volume fraction and shows a maximum at

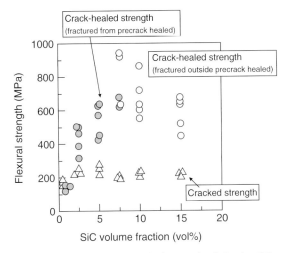

Fig. 6.7 Crack healed and cracked strength of alumina-SiC composites as a function of SiC volume fraction.

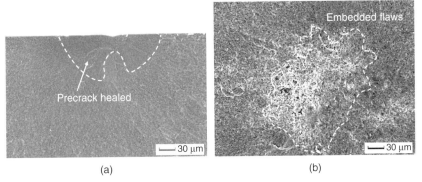

(a)   (b)

Fig. 6.8 Fracture initiations of (a) alumina/7.5 vol% SiC particles composite in which fracture initiates from the precrack healed and (b) alumina/10 vol% SiC particles composite in which fracture initiates from the embedded flaws.

SiC volume fraction of 7.5, as shown in Figure 6.7. From the strength difference, the crack-healing ability cannot be estimated by the strength recovery behavior alone. However, the fracture surface observations can reveal whether the crack is completely healed. There are two kinds of fracture mode. One is the fracture initiated from the crack-healed zone, as shown in Figure 6.8a. This means the formed oxide is not enough to heal the precrack completely. This fracture mode is observed in some specimens of the alumina mixed with 7.5 vol% SiC particles and all specimens that contain less than 5 vol% SiC particles. The other is the fracture initiated outside the crack-healed zone, as shown in Figure 6.8b. This means that the formed oxide is enough to heal the cracks. The enough quantity of the oxide is formed by the oxidation of more than 10 vol% SiC particles. From the result, the ceramics have to contain at least 10 vol% SiC to produce strong crack-healed zone.

6.5.2
SiC Figuration

SiC figuration also affects the crack-healing ability. Especially, SiC whisker that has high aspect ratio causes intrinsic change to the micromechanism of the crack healing.

Sato *et al.* [34] investigated the relation between SiC particle size and the crack-healed strength in the case of mullite $(3Al_2O_3 2SiO_2)/20$ vol% SiC composite. From the results shown in Figure 6.9, they concluded that crack healing at 1573 K for 1 h in the presence of air causes 100–300 MPa strength enhancement to all specimens, which shows a maximum with SiC particle size of 0.56 μm.

Ceramics containing SiC whiskers also show self–crack-healing ability, but there are some differences between the crack-healing behaviors driven by oxidations of SiC whiskers and that of SiC particles. This difference arises from the geometric relation between SiC whiskers and the crack wall. The SiC whiskers stick out at the crack wall and bridge between crack walls as illustrated in Figure 6.10. Owing to this geometry, partial bonding between the crack walls can be formed despite the small amount of oxide formation. The partial bonding [31, 35] was observed in the crack-healed zone of alumina/20 vol% SiC whiskers (diameter = 0.8–1.0 μm, length = 30–100 μm). The partial bonding enhances the strength recovery of crack healing and at primary stage on large cracks, as shown in Figure 6.11 [31] (cf. Figure 6.4). Both in the primary stage and on large crack, the amount of the formed oxide is too less to completely fill the crack. In this situation, large

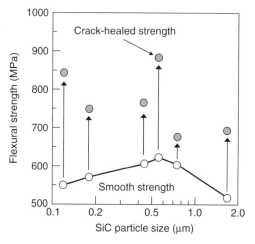

Fig. 6.9 Variation in crack-healed strength with SiC particle size in mullite containing 20 vol% SiC particles composite.

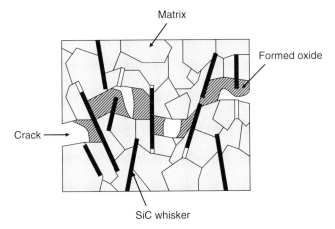

Fig. 6.10 Schematic illustration of crack-healing mechanism by SiC whiskers.

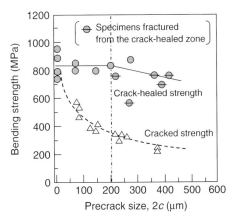

Fig. 6.11 Flexural strength of the crack-healed alumina/30 vol% SiC whiskers (diameter = 0.8–1.0 mm, length = 100 mm) composite as a function of surface length of a semi-elliptical crack.

strength recovery could not be attained without partial bonding. Therefore, composites with SiC whiskers do not only improve the fracture toughness but also have the advantage on crack-healing ability. However, the reliability of the crack-healed zone comprised by the partial bonding is inferior to that of the crack-healed zone completely filled with the formed oxide. Therefore, composites containing SiC whiskers as well as SiC particles show excellent self–crack-healing ability.

6.5.3
Matrix

Since self–crack-healing ability can only be seen in case of SiC composites, there is no restriction in selecting the matrix. Ando and coworkers succeeded to endow

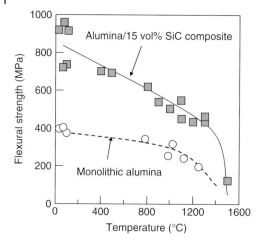

Fig. 6.12 Temperature dependence of the flexural strength of alumina containing 15 vol% SiC particles, which are entrapped in the alumina grins, compared with that of monolithic sintered alumina.

silicon nitride [24–27], alumina [28–31, 35], and mullite [21–23] with self–crack-healing ability. Also monolithic SiC [36, 37] has excellent self–crack-healing ability.

These composites can be prepared from commercially available powders using ball mill mixing and hot pressing techniques. Sintering additives does not show any influence on crack-healing ability. Moreover, a further improvement in mechanical properties can be obtained by employing the optimized sintering conditions. For example, entrapped SiC particles [28] presented in the alumina matrix grains, when alumina containing 15 vol% SiC particles composite is hot pressed at 1873 K for 4 h. The entrapped SiC particles can inhibit the glide deformation of alumina grains above 1273 K and this increases the temperature limit for bending strength, as shown in Figure 6.12 [28].

6.6
Valid Conditions

6.6.1
Atmosphere

The annealing atmosphere has an outstanding influence on the extent of crack healing and the resultant strength recovery, as shown in Figure 6.13 [29].

From the figure, it can be clearly seen that the presence of oxygen causes the self-healing phenomenon, as crack healing is driven by the oxidation of SiC. However, the threshold partial oxygen pressure can be expected to be quite low. Therefore, the self-crack healing must be valid in the atmosphere, except in

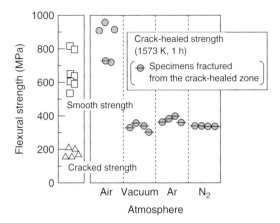

Fig. 6.13 Crack-healing behavior of alumina/15 vol% 0.27 μm SiC particles composite under several atmospheres.

deoxidized condition, for example, the atmosphere containing hydrogen. Also the embedded flaws cannot be healed, because SiC particles present in the embedded flaws cannot react with oxygen.

Furthermore, annealing in vacuum, argon (Ar), and nitrogen (N₂) results in a slight strength recovery. Wu *et al.* [20] discussed this phenomenon to be the release of the tensile residual stress at the indentation site. Furthermore, Fang *et al.* [38] used a satellite indentation technique to show that, after 2 h at 1573 K, the degree of annealing-induced relaxation in the stress intensity factor of the residual stress at the indentation site was ∼26% for alumina/5 vol% SiC nanocomposite. Using this result, one can predict that annealing in atmosphere without oxygen leads to 10% strength recovery.

6.6.2
Temperature

The ceramic components are usually operated at high temperatures. The self-healing relies on the oxidation of SiC, thereby leading to the self-crack healing. Thus, it is important to know the valid temperature range for self-crack healing.

As the crack healing is induced by chemical reaction, the strength recovery rate decreases exponentially with decreasing temperature. For example, Figure 6.14 [28] shows the relationship between crack-healing temperature and strength recovery for alumina/15 vol% 0.27 μm SiC particles composite. In order to completely heal a semi-elliptical crack of 100 μm in surface length within 1 h, heating above 1573 K is required. In the similar way, heating above 1473 and 1273 K is needed in order to completely heal the surface crack within 10 and 300 h, respectively. The relation between the crack-healing temperature and the strength recovery rate follows Arrhenius' equation.

Figure 6.15 [29] shows the Arrhenius plots on the crack healing of several ceramics having self–crack-healing ability, in which the crack-healing rate is

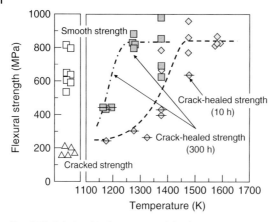

Fig. 6.14 Relationship between crack-healing temperature and strength recovery for alumina/15 vol% 0.27 μm SiC particles composite. (Centered line symbols indicate specimens fractured from the crack-healed zone.)

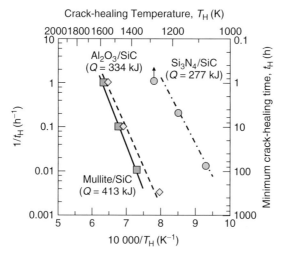

Fig. 6.15 Arrhenius plots on the crack healing of several ceramics having self-crack healing ability.

defined as the inverse of the time when complete strength recovery is attained at elevated temperatures. Apparent activation energies of crack healing can be evaluated from Figure 6.15. The question is why the activation energy of Si_3N_4/SiC composite differs from those of alumina/SiC and mullite/SiC composites. The reason could be the crack healing of Si_3N_4/SiC composite is driven by the oxidation of SiC as well as Si_3N_4. Using these values, one can estimate the time for which a semi-elliptical crack of 100 μm in surface length can be completely healed at several temperatures.

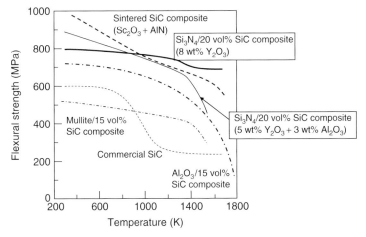

Fig. 6.16 Temperature dependences of the flexural strength of several typical crack-healed ceramics.

The refractoriness of the crack-healed zone restricts the determination of the upper limit of the valid temperature range of self-crack healing. The temperature dependence of the flexural strength of the several typical ceramics crack healed [22, 25, 28, 36, 37] is shown in Figure 6.16. Except the dependence of $Si_3N_4/20$ wt% SiC particles composite containing 8 wt% Y_2O_3 as sintering additives, all dependences of the crack-healed specimens have the temperature at which the strength decreases abruptly, and this has been determined as the temperature limit for strength. The temperature limit is affected by the features of the oxide formed by self-crack healing. The commercial sintered SiC [36] was found to have considerably low temperature limit of 873 K because the formed oxide is in glassy phase. Modifying the sintered additives to Sc_2O_3 and AlN, Lee *et al.* [37] succeeded in improving the temperature limit of the crack-healed zone significantly. The similar behavior was observed in the $Si_3N_4/20$ wt% SiC particles composites [25]. When the sintered additive is 5 wt% Y_2O_3 and 3 wt% Al_2O_3, the formed oxide and grain boundary are in glassy phase. Alternatively $Si_3N_4/20$ wt% SiC particles composite containing 8 wt% Y_2O_3 as sintering additives forms the crystalline oxide, such as $Y_2Si_2O_7$ by crack healing. The difference gives rise to the difference in the temperature limit. Both Al_2O_3 containing 15 vol% SiC particles composite [28] and mullite containing 15 vol% SiC particles composite [22] form the crystalline phase because the formed oxide reacts with the matrix and forms mullite. These temperature limits are summarized in Table 6.2.

The temperature range at which self-crack healing is valid is limited by the crack-healing rate and the high-temperature mechanical properties. Assuming that fracture by the second damage allows complete healing of surface cracks introduced by the first damage in 100 h, one can evaluate the valid temperature range of the self-crack healing, as listed in Table 6.3.

Table 6.2 Temperature limit of several ceramics.

Materials	Temperature limit (K)
Si_3N_4/20 vol% SiC particles composite (8 wt% Y_2O_3)	1573
Si_3N_4/20 vol% SiC particles composite(5 wt% Y_2O_3 + 5 wt% Al_2O_3)	1473
Alumina/15 vol% SiC particles composite	1573
Mullite/15 vol% SiC particles composite	1473
SiC sintered with Sc_2O_3 and AlN	1673
Commercial SiC sintered	873

Table 6.3 Valid temperature region of self-crack healing for several ceramics.

Materials	Valid temperature range (K)
Si_3N_4/20 vol% SiC particles composite(8 wt% Y_2O_3)	1073–1573
Alumina/15 vol% SiC particles composite	1173–1573
Mullite/15 vol% SiC particles composite	1273–1473
SiC sintered with Sc_2O_3 and AlN	1473–1673

6.6.3
Stress

Stress applied to the components is also one of the most important factors to decide the valid condition of self-crack healing. Structural components generally suffer various kinds of stresses. The applied stress is possible to cause the slow crack growth. If the applied stress exceeds the critical value, it would rise to catastrophic failure. Therefore, it is important to know the threshold stress that could be safely applied to the cracks during self-crack healing.

Ando et al. [39] have reported for the first time that the surface crack in the mullite containing 15 vol% SiC particles composite can be healed, although the tensile stress is applied to the cracks. Their results revealed that the crack healing occurs, although the precrack is grown by the applied stress, and the specimens crack healed under stress had the same strength as the specimens crack healed under no stress and at same temperature. Furthermore, Ando et al. [40] reported that surface cracks in the mullite containing 15 vol% SiC particles composite can be healed even though dynamic stress such as cyclic stress, opens and closes the crack.

Nakao et al. [41] investigated the threshold stresses during self-crack healing for several oxide ceramics. For example, the threshold static stress during crack healing for a semi-elliptical surface crack (surface length =100 μm) in alumina containing

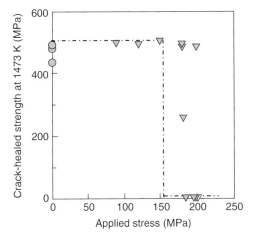

Fig. 6.17 Crack-healing behavior at 1473 K under static stress for alumina/15 vol% SiC particles composite.

15 vol% SiC particles has been determined to be 150 MPa, as shown in Figure 6.17 [41]. Figure 6.17 demonstrates that the tensile static stress of 180 MPa can fracture the specimen during crack healing, whereas the stress less than 150 MPa can never fracture the specimens during crack-healing.

Figure 6.18 shows the determined threshold stress as a function of the flexural strength of the specimen containing the same surface crack for several cracks in the oxide ceramics–SiC composite [41–44]. Except the threshold stresses of mullite containing 15 vol% SiC whiskers composite, all data satisfies the proportional relation, although the crack healings were subjected to different conditions. Crack healing ability for mullite containing 15 vol% SiC whiskers composite has been found to be so low that the crack healing part was weaker than the other parts as only partly welding occurs, not satisfying the proportional relation. The proportional constants for the relations between the threshold static and cyclic stresses between the cracked strength have been found to be 64 and 76%, respectively. The threshold stress imposes an upper limit to the crack growth rate, thereby limiting the crack length to less than the critical crack length before crack healing starts. This implied that the crack growth behavior of all specimens is time dependent rather than cyclic dependent at high temperature. Therefore, applying static stress could be confirmed to be the easiest condition for fracture during crack healing under stress, and the threshold stresses of every condition during crack healing have been found to be the threshold static stresses. The stress intensity factors at the tip of the precrack during the crack-healing treatment, K_{HS}, were estimated. Since a tensional residual stress was introduced during precracking by using the indentation method, it is necessary to consider the stress intensity factor of the residual stress, K_R, as expressed by the following equation:

$$K_{HS} = K_{ap} + K_R \qquad (6.4)$$

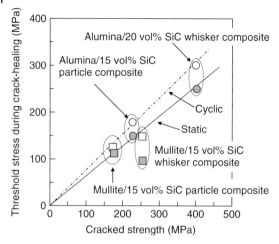

Fig. 6.18 Relation between threshold stress during crack healing and the corresponding cracked strength.

where K_R can be evaluated by using the relation proposed by Kim *et al.* [45] and $K_R = 0.35 \times K_{IC}$. Also, by interpolating the threshold static stress during crack healing and the geometry for the precrack into Newman–Raju equation [46], one can obtain K_{ap}. From the evaluation, it was found that ceramic components having the adequate crack healing ability can be crack healed under the stress intensity factor below 56% fracture toughness.

6.7
Crack-healing Effect

6.7.1
Crack-healing Effects on Fracture Probability

Crack-healing can simplify the complexity in the flaws associated with failure, because surface cracks that are severest flaws in ceramic are completely healed. As a result, a fracture probability can be easily described after crack-healing. Furthermore, crack-healing has a large contribution to decrease fracture probability.

Fracture probability is one of the most important parameters for structural components. If the fracture probability is too high, one needs to either change the design or substitute high strength materials. The fracture probability can be obtained from the failure statistics. As indicated in Section 6.2, ceramics contain many flaws that can vary in size and figuration, causing wide strength distribution. Thus the empirical approach needs to describe the strength distribution of a structural ceramic. Once the strength of a material is fitted to the distribution, the fracture probability can be predicted for any applied stress. A common empirical approach to describe the strength distribution of a structural ceramic is the Weibull approach. The two-parameter Weibull function, which is given by

$$F(\sigma) = 1 - \exp\left\{-\left(\frac{\sigma}{\beta}\right)^m\right\} \tag{6.5}$$

can express the strength distribution of structural ceramic well, where $F(\sigma)$ is the fracture probability at the tensile stress of σ, m the Weibull modulus, and β the scale parameter. The Weibull modulus describes the width of strength distribution. High Weibull modulus implies that the strength has low variability. Values of m for ceramics are in the range of 5–20. The scale parameter describes the stress when $F(\sigma) = 63.2\%$. To analyze strength distribution, Equation 6.5 is usually expanded as follows:

$$\ln\ln\left(\frac{1}{1 - F(\sigma)}\right) = m\ln\sigma - m\ln\beta \tag{6.6}$$

Thus, a plot of the left-hand side of Equation 6.6 has a linear relation versus the natural logarithm of strength. In such a procedure, a fracture probability is needed for each test specimen. This is usually estimated using

$$F(\sigma) = \frac{i - 0.3}{n + 0.4} \tag{6.7}$$

The strength data of n specimens are organized from weakest to strongest and given a rank i with $i = 1$ being the weakest specimen. Equation 6.7 is well known as *median rank method*.

As an example, the Weibull plot of the crack-healed alumina containing 20 vol% silicon carbide (SiC) particles composite is shown in Figure 6.19 [47]. The healed crack is a semi-elliptical surface crack having surface length of 100 μm and depth of 45 μm. In comparison, those of as-cracked specimen and smooth specimen having a mirror finish surface are shown in Figure 6.19. Assuming that the data obey the two-parameter Weibull function, one can apply a least-squares fitting. From the obtained line profiles, the values of m and β can be obtained for the crack-healed specimens, the as-cracked specimens and the smooth specimens.

Crack-healing causes slight increase to the value of m compared to the smooth specimens. Furthermore, the strength distribution of the crack-healed specimen is in good agreement with the two-parameter Weibull function, although that of the smooth specimen differs from the function significantly. The flaw population in ceramics leads to this behavior. The Weibull modulus m of the crack-healed specimen was smaller than that of the as-cracked specimen. All as-cracked specimens fractured from a crack introduced by the Vickers indentation, while fractures of most crack-healed specimens occurred outside of the crack-healed zone, as shown in Figure 6.3b, because cracks were completely healed. Since the embedded flaws

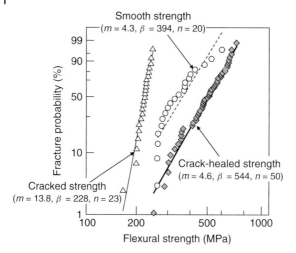

Fig. 6.19 Weibull plot of the crack-healed alumina containing 20 vol% silicon carbide (SiC) particles composite.

as the fracture initiation of the crack-healed specimens have different sizes, the fracture stresses exhibit a large scatter.

This specimen was tested at room temperature and exhibited the fracture stress of 526 MPa. To improve the reliability and the quality of the structural ceramics, it is therefore necessary to remove the specimens with large embedded flaws by proof test, even if surface cracks were completely healed. The scale parameter of the crack-healed specimens has a higher value than that of as-cracked specimens as well as that of the smooth specimens, because cracks introduced by machining, which existed even in the smooth specimens, were also completely healed.

To show the considerable merit of crack healing, the fracture probabilities of three specimens (smooth specimens, as-cracked specimens, and crack-healed specimens) for the proof test stress of 435 MPa were compared. The fracture probabilities of smooth specimens, as-cracked specimens, and crack-healed specimens were 80, 100, and 30%, respectively. Therefore, it can be concluded that crack healing drastically increases the survival probability by proof test, and thus increases the working stress of the structural ceramics.

6.7.2
Fatigue Strength

The effect of self-crack healing on the fatigue strength is greater than that on the monotonic strength. The fatigue degradation of ceramics progresses by the stress corrosion cracking at the tip of the crack, as mentioned in Section 6.2. Therefore, the presence of surface cracks affects strongly the fatigue strength. Figure 6.20 [48] shows the dynamic fatigue results of the crack-healed mullite containing 15 vol% SiC whiskers and 10 vol% SiC particles composite

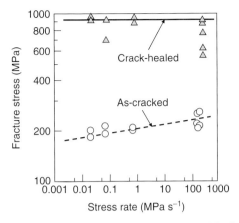

Fig. 6.20 Dynamic fatigue results of the crack-healed mullite containing 15 vol% SiC whiskers and 10 vol% SiC particles composite with that of the composite having a semi-elliptical crack of 100 μm in surface length.

with that of the composite having a semi-elliptical crack of 100 μm in surface length. From logarithmic plots of the dynamic strength versus stressing rate, the effect of crack healing on the fatigue behavior can be clearly understood. The positive slope implies that the slow crack growth occurs. As the data on the specimens containing the surface crack shows the positive slope, the slow crack growth has been included in the fatigue behavior. On the other hand, the data on the crack-healed specimen is almost constant over the whole stressing rate. Therefore, crack healing makes the fatigue sensitivity decrease significantly. Actually, the fracture initiator in the crack-healed mullite containing 15 vol% SiC whiskers and 10 vol% SiC particles composite under the every stressing rate is an embedded flaw, which did not grow under the applied stress.

In high temperature fatigue, there is another interesting phenomenon, in order that self-crack healing occurs at the same time as fatigue damage. For example, Figure 6.21 [48] shows a logarithmic plot of life time in terms of the applied stress for the crack-healed mullite containing 15 vol% SiC whiskers and 10 vol% SiC particles composite at 1273 K. In general, that is, the slow crack growth is included, the life time increases as the applied stress decreases. However, all crack-healed test specimens survived up to the finish time of 100 h obeying the JIS standard [49] under static stresses of 50 MPa; lesser than the lower limit of the monotonic strength at the same temperature. Alternatively, the specimens fractured at less than 100 s under stresses corresponding to the lower limit of the flexural strength. This failure is not fatigue, but rather rapid fracture. Therefore, it is confirmed that the crack-healed composite is not degraded by the static fatigue at 1273 K. The behavior would result from the fact that self-crack healing occurs rapidly compared with the fatigue damage.

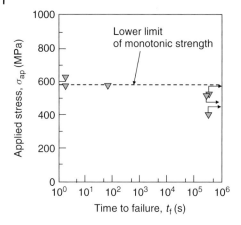

Fig. 6.21 Logarithmic plot of life time in terms of the applied stress for the crack-healed mullite containing 15 vol% SiC whiskers and 10 vol% SiC particles composite at 1273 K.

6.7.3
Crack-healing Effects on Machining Efficiency

An important alternative aspect is that self-crack healing is a most valuable surface treatment. Applying the crack-healing process into the manufacturing for the ceramic component can reduce the manufacturing cost. Machining process included in the manufacturing reduces the reliability of the components [50] because it causes many cracks to the surface of the component. To remove the nonacceptable cracks, final machining processes, such as polishing and lapping, are generally required. Although these processes leave behind many minute cracks, these are expensive processes. It is, therefore, anticipated that substituting the crack-healing process for the final machining processes leads to economical manufacturing for ceramic components secured with high reliability.

Figure 6.22 [51] demonstrates that the nonacceptable cracks introduced by heavy machining can be completely crack healed. The machining cracks were introduced at the bar specimens surface of alumina/20 vol% SiC whiskers composite by ball-drill grinding. The ball-drill grinding was performed along the direction perpendicular to the long side of the specimens, as shown in shown in Figure 6.6, which consequently fabricated a semi-circular groove whose depth and curvature were 0.5 and 2 mm, respectively. As a result, the machined specimens contained many machining cracks perpendicular to the tensile stress at the bottom of the semi-circular groove. The horizontal variable is cut depth by one pass, which is an indicator of the machining efficiency. For example, 14 to 40 cycles are needed to fabricate the semi-circular groove by the grinding with the cut depth by one pass of 15 and 5 µm, respectively. Alternatively, the vertical axis indicates the local fracture stress of the machined specimens and machined specimen healed. From the load as specimens fractured, P_F, section modulus, Z, and stress concentration

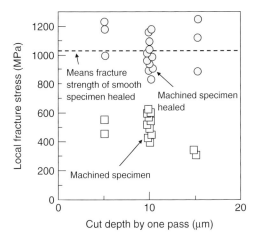

Fig. 6.22 Effect of depths of cut by one pass on the local fracture stress at room temperature of the machined specimens healed.

factor, α, the local fracture stress at the bottom of the semi-circular groove, σ_{LF}, was evaluated as follows:

$$\sigma_{LF} = \frac{\alpha P_F l}{4Z} \tag{6.8}$$

where l is span length. Under this geometry, the values of Z and α were 4.2 and 1.4 [52], respectively. The local fracture stress of machined specimens decreased with increasing cutting depth. This behavior implies that the cut depth by one pass also means the degree of the machining heaviness. Throughout the range of the cut depth by one pass, complete strength recovery was found to be attained by crack-healing treatment for 10 h at 1673 K, because these average strengths were almost equal to that of the complete crack-healed specimens [35]. Crack healing is possible for relatively large cracks initiated by heavy machining for cutting depths up to 15 μm. However, heavy machining for cutting depths above 15 μm makes the diamond grain to drop out of the ball drill significantly, reducing the machining efficiency. Therefore, with a simple operation of heating, one can ensure the reliability over ceramic components machined by the limiting conditions of grinding tool (ball-drill). To not cause the outstanding strength decrease by fabricating the semi-circular groove, it is necessary to perform not only on machining with cut depth by one pass of less than 5 μm but also on lapping at the bottom. Thus, high machining efficiency can be attained by the use of the crack-healing process.

It is important to note the difference in the optimized condition between crack healing for indented cracks and machining cracks. The optimized crack healing condition for indented cracks in alumina/20 vol% SiC whiskers composite was found to be 1573 K for 1 h [51]. However, crack healing above 1673 K for more than 10 h is required to attain the complete strength recovery for machined alumina/20

Table 6.4 Weibull modulus (*m*) and shape parameter (β) of ball-grind alumina composites containing 20 vol% of SiC whiskers.

Sample type	Weibull modulus, *m* (MPa)	Shape parameter (β)
Machined specimen	549	6.69
Machined specimen healed at 1573 K for 1 h	796	6.70
Machined specimen healed at 1673 K for 10 h	1026	11.5
Healed (1573 K for 1 h) smooth specimen	1075	8.15

vol% SiC whiskers composite. This behavior can be clearly understood by the statistical analysis using the two-parameter Weibull function given by Equation 6.5. Table 6.4 shows the scale parameter and the shape parameter evaluated from the analyses. The values of *m* and β of the machined specimen healed at 1673 K for 10 h were 1026 MPa and 11.5, respectively. The values were almost equal to those of complete crack-healed specimen. From this statistical and χ^2 analysis, it was found that the cracks introduced by machining were completely healed by the crack healing process at 1673 K for 10 h. On the other hand, the machined specimen healed at 1573 K for 1 h had lower values of *m* and β than those of the smooth specimen healed; thus, it can be concluded that this crack-healing condition is inadequate. Two approaches are confirmed to be reasonable to explain this difference: (i) the difference in the state of the subsurface residual stress associated with the different crack geometries and (ii) the oxidation of SiC by the heat generation during machining. The machining crack was closed by the action of the compressive residual stress resulting in a reduction in the supply of oxygen to the crack surfaces. Moreover, before the crack-healing

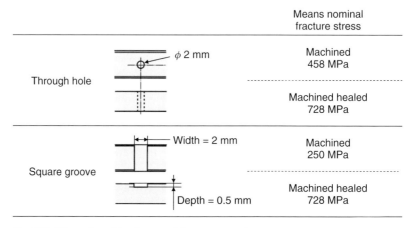

Fig. 6.23 Effect of crack-healing condition on strength recoveries of various machined specimens.

treatment, if SiC particles were already covered with a thin oxidation layer, by the grinding heat, this would lead to the decrease in the oxidation activity of SiC particles.

Furthermore, it was found that the various cracks initiated by machining into various machined figurations could be healed by crack-healing treatment at 1673 K for 10 h, as shown in Figure 6.23 [51].

6.8
New Structural Integrity Method

6.8.1
Outline

A combination of the crack healing and proof testing ultimately guarantees the structural integrity of the ceramic components. As mentioned above, by eliminating not only the cracks introduced during manufacturing but also the cracks introduced during service, the crack healing can ensure perfectly against risk of the fracture initiated from the surface flaws. The proof testing, in which components are over-stressed prior to use, can determine the maximum critical size of the embedded flaws associated with fracture. The embedded flaws, such as voids, may be present in a material as a result of the processing, and could not be generated during service. Therefore, the minimum fracture stress caused by the embedded flaws [53, 54] or probabilistic fatigue S−N curves [55−57] has been estimated from the proof testing stress on the basis of linear fracture mechanics.

However, two important points must be taken account of while using the proof testing. First is that engineering ceramics show nonlinear fracture behavior [58, 59]. The other is that the evaluated minimum fracture stress is valid only at proof testing temperature. Therefore, if the ceramic components are used at high temperature, the proof testing must be conducted at the operating temperature. Ando *et al.* [60] have proposed a theory to evaluate the temperature dependence of the guaranteed (minimum) fracture stress of a proof tested sample based on nonlinear fracture mechanics. This approach allows the ceramic components proof tested at room temperature to operate at the arbitrary temperatures.

6.8.2
Theory

On the basis of the process zone size failure criterion, the minimum fracture stress at high temperature can be guaranteed from the proof testing stress at room temperature. The criterion proposed by Ando *et al.* [60] has been obtained by the size of process zone, which is the plastic deformation region at the crack tip, well expressing the nonlinear fracture behavior of ceramics. From the criterion, the process zone size at failure, D_C, is given by the following equation:

$$D_C = \frac{\pi}{8}\left(\frac{K_{IC}}{\sigma_0}\right)^2 = a_e \left\{\sec\left(\frac{\pi\sigma_C}{2\sigma_0}\right) - 1\right\} \tag{6.9}$$

where σ_C and σ_0 are the fracture stress of the fracture strength caused by the flaws having a_e and the plain specimen (intrinsic bending strength), respectively, K_{IC} plane strain fracture toughness and a_e equivalent crack length, which is given by the equation as

$$K_C = \sigma_C\sqrt{\pi a_e} \tag{6.10}$$

By using Equation (6.9) to arrive at a_e, one can write the equivalent crack size of the flaws associated with fracture as shown in the following equation:

$$a_e = \frac{\pi}{8}\left(\frac{K_{IC}}{\sigma_0}\right)^2 \left\{\sec\left(\frac{\pi\sigma_C}{2\sigma_0}\right) - 1\right\}^{-1} \tag{6.11}$$

Since K_{IC} and σ_0 is the function of temperature, the fracture strength associated with a_e needs to be also expressed as a function of temperature:

$$\sigma_C^T = \frac{2\sigma_0^T}{\pi}\arccos\left\{\frac{\pi}{8}\frac{1}{a_e}\left(\frac{K_{IC}^T}{\sigma_0^T}\right)^2 + 1\right\}^{-1} \tag{6.12}$$

where superscript T is the value at elevated temperature.

The maximum size of the flaw, a_e^P, that is able to present in the sample proof tested under σ_p at room temperature can be given by

$$a_e^P = \frac{\pi}{8}\left(\frac{K_{IC}^R}{\sigma_0^R}\right)^2 \left\{\sec\left(\frac{\pi\sigma_p}{2\sigma_0^R}\right) - 1\right\}^{-1} \tag{6.13}$$

where superscript R is the value at room temperature. Assuming that the sizes of the residual embedded flaws do not vary with change in the temperature, one can determine the minimum fracture stress guaranteed of the proof tested sample, σ_G, as the fracture strength associated with a_e^P at arbitrary temperatures. Thus the value of σ_G can be expressed as

$$\sigma_G = \frac{2\sigma_0^T}{\pi}\arccos\left\{\left(\frac{K_{IC}^T}{K_{IC}^R}\right)^2\left(\frac{\sigma_0^R}{\sigma_0^T}\right)^2\left\{\sec\left(\frac{\pi\sigma_p}{2\sigma_0^R}\right) - 1\right\} + 1\right\}^{-1} \tag{6.14}$$

and can be evaluated from the data on the temperature dependences of K_{IC} and σ_0.

6.8.3
Temperature Dependence of the Minimum Fracture Stress Guaranteed

Using the above theory, one can estimate the minimum fracture stresses at elevated temperatures for the sample proof tested at room temperature. Ono *et al.* [47] evaluated the temperature dependence of σ_G for alumina/20 vol% SiC particles composite. Moreover, by comparing the evaluated σ_G with the measured fracture stress of the proof tested sample at elevated temperature, the validity of this estimation was given by them. Their results and discussion are presented in the following text.

Before discussing the temperature dependence of σ_G, temperature dependences of the plane strain fracture toughness, K_{IC}, and the intrinsic bending strength, σ_0, are noticed. Ono *et al.* [47] investigated these temperature dependences for alumina/20 vol% SiC particles composite. The σ_0, which is determined as the average fracture stress of 5% of the highest strengths of the crack-healed specimens at the temperatures, has large temperature dependence and the tendency is almost linear and negative up to 1373 K. Moreover, the K_{IC} is almost constant against temperature. The features affect the correlativeness between the fracture stresses as a function of the equivalent crack length, a_e, at room temperature and at high temperature. The schematic is shown in Figure 6.24.

The fracture stress associated with small flaw, that is, a_e is low, is equal to the σ_0, and varies considerably as temperature varies. On the other hand, the fracture stress associated with large flaw, that is, a_e is high, is determined by linear fracture mechanics as expressed by Equation 6.10, and changes scarcely with temperature change. Therefore, the negative temperature gradient of the σ_G increases with increasing proof testing stress, σ_p^R, as shown in Figure 6.25. Since high σ_p^R qualifies a_e^p to be a low value, the negative temperature gradient of the σ_G becomes considerably high. Alternatively, since low σ_p^R allows the large flaws to

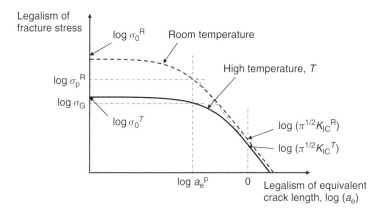

Fig. 6.24 Schematic illustration of proof-test theory and the effect of equivalent crack on fracture strength at room temperature and at high temperature.

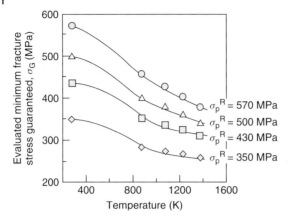

Fig. 6.25 Temperature dependence of minimum fracture stress guaranteed of the proof tested under several proof testing stress for alumina containing 20 vol% SiC particles composite.

present in the proof tested specimen, the σ_G is almost constant as a function of temperature.

The values of the evaluated σ_G have good agreements with the measured minimum fracture stress of the proof tested specimens, σ_F^{min}. Figure 6.26 shows the data on fracture stress of the crack-healed and proof-tested specimens as a function of temperature with the evaluated σ_G for the crack-healed alumina/20 vol% SiC particles composite when $\sigma_p = 435\,MPa$. Except the data at 1373 K, all specimens have higher strength than the σ_G at all the temperature.

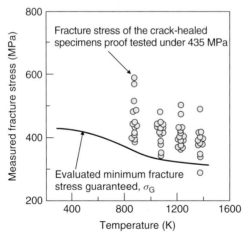

Fig. 6.26 Comparison between measured fracture stress and the evaluated minimum fracture stress guaranteed for the crack-healed alumina/20 vol% SiC particles composite proof tested under 435 MPa.

Also, the minimum values of the experimental fracture stress are almost equal to the σ_G at all temperatures. At 1373 K, the σ_F^{min} is 6.8% less than σ_G, but the value exists in the dispersion evaluated from the K_{IC} and the σ_0 that have large scatters. Moreover, in the case of $\sigma_p = 530$ MPa, the evaluated σ_G have good agreements with the σ_F^{min} as well as with all the proof-tested specimens fractured under the tensile stress more than σ_G. Therefore, the results well demonstrate the validity of the guaranteed method.

Moreover, Ono *et al.* [47] and Ando *et al.* [60, 61] reported that the guaranteed theory can be applied to different conditions and different materials. The obtained results can be seen in Figure 6.27, where the measured σ_F^{min} is plotted as a function of the evaluated σ_G. N in the figure denotes the number of samples used to obtain σ_F^{min}. Four open diamonds indicate the data on the ceramic coil spring made of silicon nitride. Also, a open square differs from the closed circles in the crack healing condition, that is, open square employs 1373 K for 50 h and closed circle employs 1573 K for 1 h. However, all σ_F^{min} shows good agreement with σ_G. Therefore, using Equation 6.14, one can estimate the σ_G at higher temperatures of every material having crack healing ability and every crack healing condition.

On the other point of view, it is interesting whether this estimation is reversible for temperatures, that is, the stress evaluated in Equation 6.14 can guarantee the minimum fracture stress at room temperature of the specimen proof tested at high temperature. For example, the σ_G at room temperature of the alumina/20 vol% SiC particles composite proof tested under 335 MPa ($= \sigma_p$) at 1073 K is evaluated to be 435 MPa. Alternatively, the experimental minimum fracture stress was 410 MPa. The evaluated σ_G existed in the dispersion obtained from K_{IC} and σ_0, which have large scatters.

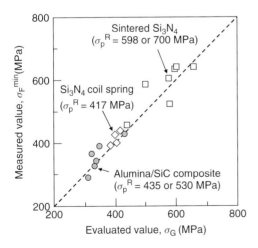

Fig. 6.27 Comparison between minimum fracture stresses guaranteed and measured minimum fracture stress.

6.9
Advanced Self-crack Healing Ceramics

6.9.1
Multicomposite

Ceramic composite containing both SiC whiskers and SiC particles, called *SiC multicomposite*, enhances fracture strength and toughness as well as endows the ceramics with good self-crack healing ability. As mentioned in Section 6.5, reinforcement by SiC whiskers can not only improve fracture toughness but can also generate self-crack healing ability. However, it is difficult to disperse large amount of SiC whiskers uniformly, and so the reinforcement of aggregated SiC whiskers decreases the fracture strength. SiC multicomposite containing both whiskers and particles improves the self-crack healing ability endowed by SiC whiskers alone without adversely affecting the composite strength. Therefore, ceramic–SiC multicomposites exhibit high strength, high fracture toughness, and excellent self–crack-healing ability.

Especially, SiC multicomposites perform better than the mullite-based composites. Mullite and mullite-based composites have been expected to be advanced ceramic spring because they exhibit the same low level elastic constant as metal and excellent oxidation resistance. However, mullite has remarkably low fracture toughness. Therefore, it is necessary to endow mullite with self–crack- healing ability for actualizing mullite-based ceramic springs. The mechanical properties of mullite/SiC composites [21, 44] and multicomposites [62] were investigated as shown in Table 6.5.

Fracture strength increases with SiC content increasing up to 20 vol%; above which it remains almost constant. Fracture toughness increased with an increase in SiC whiskers content. Clearly, it is confirmed that crack bridging and pulling out due to SiC whiskers lead to increase in fracture toughness. All mullite/SiC composites, which are listed in Table 6.5, can exhibit large strength recovery by crack healing. However, mullite/15 vol% SiC whiskers composite (MS15W) [44]

Table 6.5 Mullite/SiC composites having self-crack healing ability.

Sample descriptions	Content (vol%)		
	Mullite	SiC particle (diameter $= 0.27\,\mu m$)	SiC whisker (diameter $= 0.8 - 1.0\,\mu m$, length $= 30 - 100\,\mu m$)
MS15P	85	15	0
MS15W	85	0	15
MS20W	80	0	20
MS25W	75	0	25
MS15W5P	80	5	15
MS15W10P	75	10	15

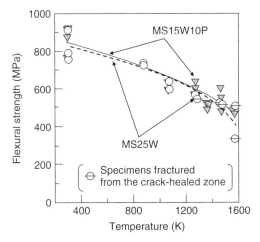

Fig. 6.28 Temperature dependence of the crack-healed strength of mullite/25 vol% SiC whiskers composite (MS25W) and mullite/15 vol% SiC whiskers/10 vol% SiC particles composite (MS15W10P), in which the center lined symbols indicate specimens fractured from the precrack healed.

cannot attain the complete strength recovery despite the optimized crack healing condition. Complete strength recovery of the crack-healed specimens is defined as the strength of the specimens, whose fracture initiation is embedded flaw. This implies complete elimination of surface cracks that can be attained and the strength of the crack healed part is superior to that of the base materials. The crack healed part in mullite/25 vol% SiC whiskers composite (MS25W) holds higher strength than the base materials below 1273 K, as shown in Figure 6.28 [62]. On the other hand, the crack healed part in mullite/15 vol% SiC whiskers/10 vol% SiC particles composite (MS15W10P) holds higher strength than the base materials at the whole of the experimental temperature region, as shown in Figure 6.28.

Figure 6.29 [62] shows the maximum shear strains of the mullite/SiC multicomposites as a function of SiC content. The maximum shear strain corresponds to the deformation ability as spring. The value of the maximum shear strain showed a maximum at a SiC content of 20 vol%, above which it slightly decreased because Young's modulus increased with an increase in SiC content, but the fracture strengths were almost constant above SiC content of 20 vol%. MS15W10P has the best potential as a material for ceramic spring used at high temperatures, because it has a shear deformation ability that was almost two times greater than monolithic mullite as well as an adequate crack healing ability.

6.9.2
SiC Nanoparticle Composites

Nanometer-sized SiC fine particles enhance the self-crack healing rate because it gives large increment in reactive area and makes the surface of SiC particles active.

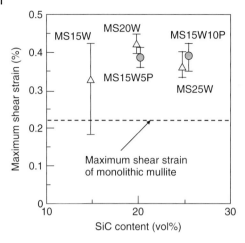

Fig. 6.29 Maximum shear strains of mullite/SiC composites as a function of SiC content.

This effect gives large benefit to self-crack healing at relatively low temperatures at which self-crack healing is completed in more than 100 h.

Reaction synthesis is a promising process for directly fabricating nanocomposites which are difficult to obtain by the normal sintering of nanometer-sized starting powder compacts. Reaction synthesis to fabricate alumina-SiC nanocomposite [63–69] were reported. Using the reaction synthesis (6.15)

$$3(3Al_2O_3 2SiO_2) + 8Al + 6C = 13Al_2O_3 + 6SiC \qquad (6.15)$$

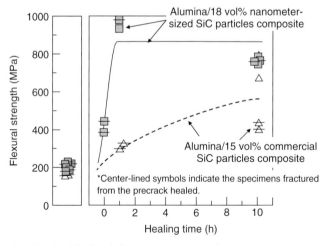

Fig. 6.30 Crack-healing behavior at 1373 K on alumina containing 18 vol% nanometer-sized SiC particles composite and alumina containing 15 vol% commercial SiC particles composite, in which center lined symbols indicate the specimen fractured from the precrack healed.

Zhang *et al.* [69] succeeded in fabricating alumina nanometer-sized SiC particles nanocomposite, in which the formed SiC particles are mainly entrapped inside the alumina grains. Employing the similar process to prepare alumina–SiC nanocomposite, Nakao *et al.* [70] investigated the effect of nanometer-sized SiC particle on the crack-healing behavior, as shown in Figure 6.30.

The result demonstrates that the nanometer-sized SiC with particle size of 20 nm can significantly increases the self-crack healing rate compared to the commercial 270 nm SiC particles. Furthermore, the nano-SiC particles can attain the complete strength recovery within 10 h at 250 K, which is a lower temperature compared to the commercial SiC particles. Although nano-SiC makes crack-healing reaction activated at lower temperatures, it gives same level of refractoriness as the alumina containing commercial SiC particles composite. Therefore, it is noted that the use of SiC nanoparticles is a most valuable route to enhance the valid temperature region of self crack healing.

References

1 Williams, L.S. (**1956**) *Transactions of the British Ceramic Society*, **55**(5), 287–312.

2 Evans, A.G. (**1972**) *Journal of Materials Science*, **7**(10), 1131–46.

3 Dwivedi, P.J. and Green, D.J. (**1995**) *Journal of the American Ceramic Society*, **78**(8), 2122–28.

4 Heuer, A.H. and Roberts, J.P. (**1966**) *Proceedings of the British Ceramic Society*, **6**, 17–27.

5 Lange, F.F. and Gupta, T.K. (**1970**) *Journal of the American Ceramic Society*, **53**(1), 54–55.

6 Davies, L.M. (**1966**) *Proceedings of the British Ceramic Society*, **6**, 29–53.

7 Lange, F.F. and Radford, K.C. (**1970**) *Journal of the American Ceramic Society*, **53**(7), 420–21.

8 Roberts, J.T.A. and Wrona, B.J. (**1973**) *Journal of the American Ceramic Society*, **56**(6), 297–99.

9 Bandyopadhyay, G. and Roberts, J.T.A. (**1976**) *Journal of the American Ceramic Society*, **59**(9-10), 415–19.

10 Gupta, T.K. (**1976**) *Journal of the American Ceramic Society*, **59**(9-10), 448–49.

11 Evans, A.G. and Charles, E.A. (**1977**) *Acta Metallurgica*, **25**, 919–27.

12 Lange, F.F. (**1970**) *Journal of the American Ceramic Society*, **53**(5), 290.

13 Easler, T.E., Bradt, R.C. and Tressler, R.E. (**1982**) *Journal of the American Ceramic Society*, **65**(6), 317–20.

14 Chu, M.C., Cho, S.J., Yoon, K.J. and Park, H.M. (**2005**) *Journal of the American Ceramic Society*, **88**(2), 491–93.

15 Niihara, K. and Nakahira, A. (**1998**) "Strengthening of oxide ceramics by SiC and Si_3N_4 dispersions", *Proceeding of the Third International Symposium on Ceramic Materials and Components for Engines*, American Ceramics Society, Westerville, pp. 919–26.

16 Niihara, K. (**1991**) *Journal of the Ceramic Society of Japan*, **9**(10), 974–82.

17 Niihara, K., Nakahira, A. and Sekino, T. (**1993**) *Materials Research Society Symposium Proceedings*, **286**, 405–12.

18 Thompson, A.M., Chan, H.M. and Harmer, M.P. (**1995**) *Journal of the American Ceramic Society*, **78**(3), 567–71.

19 Chou, I.A., Chan, H.M. and Harmer, M.P. (**1998**) *Journal of the American Ceramic Society*, **81**(5), 1203–8.

20 Wu, H.Z., Lawrence, C.W., Roberts, S.G. and Derby, B. (**1998**) *Acta Materialia*, **46**(11), 3839–48.

21 Chu, M.C., Sato, S., Kobayashi, Y. and Ando, K. (**1995**) *Fatigue and*

Fracture of Engineering Materials and Structures, **18**(9), 1019–29.

22 Ando, K., Tsuji, K., Hirasawa, T., Kobayashi, Y., Chu, M.C. and Sato, S. (**1999**) *Journal of the Society of Materials Science, Japan*, **48**(5), 489–94.

23 Ando, K., Tsuji, K., Ariga, M. and Sato, S. (**1999**) *Journal of the Society of Materials Science, Japan*, **48**(10), 1173–78.

24 Ando, K., Ikeda, T., Sato, S., Yao, F. and Kobayasi, Y. (**1998**) *Fatigue and Fracture of Engineering Materials and Structures*, **21**, 119–22.

25 Ando, K., Chu, M.C., Kobayashi, Y., Yao, F. and Sato, S. (**1991**) *The Japan Society of Mechanical Engineering, International Journal Series A*, **65 A**, 1132–39.

26 Ando, K., Chu, M.C., Yao, F. and Sato, S. (**1999**) *Fatigue and Fracture of Engineering Materials and Structures*, **22**, 897–903.

27 Yao, F., Ando, K., Chu, M.C. and Sato, S. (**2001**) *Journal of the European Ceramic Society*, **21**, 991–97.

28 Ando, K., Kim, B.S., Chu, M.C., Saito, S. and Takahashi, K. (**2004**) *Fatigue and Fracture of Engineering Materials and Structures*, **27**, 533–41.

29 Kim, B.S., Ando, K., Chu, M.C. and Saito, S. (**2003**) *Journal of the Society of Materials Science, Japan*, **52**(6), 667–73.

30 Ando, K., Kim, B.S., Kodama, S., Ryu, S.H., Takahashi, K. and Saito, S. (**2003**) *Journal of the Society of Materials Science, Japan*, **52**(11), 1464–70.

31 Nakao, W., Osada, T., Yamane, K., Takahashi, K. and Ando, K. (**2005**) *Journal of the Japan Institute of Metals*, **69**(8), 663–66.

32 Chase M.W. Jr. (ed.) (**1998**) *NIST-JANAF Thermochemical Tables*, 4th edn, American Chemistry Society and American Institute of Physics for the National Institute of Standards and Technology.

33 Japan Industrial Standard R1601 (**1993**) *Testing Method for Flexural Strength of High Performance Ceramics*, Japan Standard Association, Tokyo.

34 Sato, S., Chu, M.C., Kobayashi, Y. and Ando, K. (**1995**) *Journal of the Japan Society of Mechanical Engineers*, **61**, 1023–30.

35 Takahashi, K., Yokouchi, M., Lee, S.K. and Ando, K. (**2003**) *Journal of the American Ceramic Society*, **86**(12), 2143–47.

36 Lee, S.K., Ishida, W., Lee, S.Y., Nam, K.W. and Ando, K. (**2005**) *Journal of the European Ceramic Society*, **25**(5), 569–76.

37 Lee, S.K., Ando, K. and Kim, Y.W. (**2005**) *Journal of the American Ceramic Society*, **88**(12), 3478–82.

38 Fang, J., Chan, H.M. and Harmer, M.P. (**1995**) *Materials Science and Engineering A*, **195**, 163–67.

39 Ando, K., Furusawa, K., Chu, M.C., Hanagata, T., Tuji, K. and Sato, S. (**2001**) *Journal of the American Ceramic Society*, **84**(9), 2073–78.

40 Ando, K., Furusawa, K., Takahashi, K., Chu, M.C. and Sato, S. (**2002**) *Journal of the Ceramic Society of Japan*, **110**(8), 741–47.

41 Nakao, W., Takahashi, K. and Ando, K. (**2006**) *Materials Letters*, **61**, 2711–13.

42 Ando, K., Yokouchi, M., Lee, S.K., Takahashi, K., Nakao, W. and Suenaga, H. (**2004**) *Journal of the Society of Materials Science, Japan*, **53**(6), 599–606.

43 Nakao, W., Ono, M., Lee, S.K., Takahashi, K. and Ando, K. (**2005**) *Journal of the European Ceramic Society*, **25**(16), 3649–55.

44 Ono, M., Ishida, W., Nakao, W., Ando, K., Mori, S. and Yokouchi, M. (**2004**) *Journal of the Society of Materials Science, Japan*, **54**(2), 207–14.

45 Kim, B.A., Meguro, S., Ando, K. and Ogura, N. (**1990**) *Journal of the High Pressure Institute of Japan*, **28**, 218–23.

46 Newman, J.C. and Raju, I.S. (**1981**) *Engineering Fracture Mechanics*, **15**, 185–92.

47 Ono, M., Nakao, W., Takahashi, K., Nakatani, M. and Ando, K. (**2007**) *Fatigue and Fracture of*

Engineering Materials and Structures, **30**(7), 599–667.

48 Nakao, W., Nakamura, J., Yokouchi, M., Takahashi, K. and Ando, K. (**2006**) *Transactions of JSSE*, **51**, 20–26.

49 Japan Industrial Standard R1632 (**1998**) *Test Method for Static Bending Fatigue of Fine Ceramics*, Japan Standard Association, Tokyo.

50 Kanematsu, W., Yamauchi, Y., Ohji, T., Ito, S. and Kubo, K. (**1992**) *Journal of the Ceramic Society of Japan*, **100**(6), 775–79.

51 Osada, T., Nakao, W., Takahashi, K., Ando, K. and Saito, S. (**2007**) *Journal of the Ceramic Society of Japan*, **115**(4), 278–84.

52 Nishida, M. (**1967**) *Stress Concentration*, Morikita Publishing, Tokyo, pp. 572–74.

53 Ritter J.E. Jr., Oates, P.B., Fuller E.R. Jr. and Wiederhorn, S.M. (**1980**) *Journal of Materials Science*, **15**, 2275–81.

54 Ritter J.E. Jr., Oates, P.B., Fuller, E.R. Jr. and Wiederhorn, S.M. (**1980**) *Journal of Materials Science*, **15**, 2282–95.

55 Hoshide, T., Sato, T. and Inoue, T. (**1990**) *Journal of the Japan Society of Mechanical Engineers A*, **56**, 212–18.

56 Hoshide, T., Sato, T., Ohara, T. and Inoue, T. (**1990**) *Journal of the Japan Society of Mechanical Engineers A*, **56**, 220–23.

57 Ando, K., Sato, S., Sone, S. and Kobayashi, Y. "Probabilistic study on fatigue life of proof tested ceramics spring", in *Fracture From Defects, Proceedings of ECF-12* (eds M.W. Brown, E.R. de los Rios and K.J. Miller), EMAS Publishing, London, **1998**, pp. 569–74.

58 Ando, K., Kim, B.A., Iwasa, M. and Ogura, N. (**1992**) *Fatigue and Fracture of Engineering Materials and Structures*, **15**, 139–49.

59 Ando, K., Iwasa, M., Kim, B.A., Chu, M.C. and Sato, S. (**1993**) *Fatigue und Fracture of Engineering Materials and Structures*, **16**, 995–1006.

60 Ando, K., Shirai, Y., Nakatani, M., Kobayashi, Y. and Sato, S. (**2002**) *Journal of the European Ceramic Society*, **22**, 121–28.

61 Ando, K., Takahashi, K., Murase, H. and Sato, S. (**2003**) *Journal of the High Pressure Institute of Japan*, **41**, 316–26.

62 Nakao, W., Mori, S., Nakamura, J., Yokouchi, M., Takahashi, K. and Ando, K. (**2006**) *Journal of the American Ceramic Society*, **89**(4), 1352–57.

63 Chaklader, A.C.D., Gupta, S.D., Lin, E.C.Y. and Gutowski, B. (**1992**) *Journal of the American Ceramic Society*, **75**(8), 2283–85.

64 Borsa, C.E., Spiandorello, F.M. and Kiminami, R.H.G.A. (**1999**) *Materials Science Forum*, **299-300**, 57–62.

65 Amroune, A., Fantozzi, G., Dubois, J., Deloume, J.P., Durand, B. and Halimi, R. (**2000**) *Materials Science and Engineering A*, **290**, 11–15.

66 Amroune, A. and Fantozzi, G. (**2001**) *Journal of Materials Research*, **16**, 1609–13.

67 Lee, J.H., An, C.Y., Won, C.W., Cho, S.S. and Chun, B.S. (**2000**) *Materials Research Bulletin*, **35**, 945–54.

68 Pathank, L.C., Bandyopadhyay, D., Srikanth, S., Das, S.K. and Ramachandrarao, P. (**2001**) *Journal of the American Ceramic Society*, **84**(5), 915–20.

69 Zhang, G.J., Yang, J.F., Ando, M. and Ohji, T. (**2004**) *Journal of the American Ceramic Society*, **87**(2), 299–301.

70 Nakao, W., Tsutagawa, Y. and Ando, K. (**2008**) *Journal of Intelligent Material Systems and Structures*, **19**, 407–10.

7

Self-healing of Metallic Materials: Self-healing of Creep Cavity and Fatigue Cavity/crack

Norio Shinya

7.1
Introduction

Metallic materials have been thought to be sufficiently reliable and secure, and rarely cause failures in their structures, owing to their high strength/toughness and flawless structures. Long-term service, however, in severe corrosive, fatigue, and creep environments causes unexpected sudden fractures of the materials due to accumulated damages. The material fracture would lead to failures of industrial plants and public infrastructures and sometimes disasters affecting people's lives. The creep cavity and fatigue cavity/crack leading to creep and fatigue fractures are too fine to be detected by nondestructive tests, and also difficult to be repaired on the service sites. The creep cavity and fatigue cavity/crack, therefore, are the damages that should be self-healed autonomously during service of the materials. Recent progresses in self-healings and coatings with respect to corrosion damages are reported in other chapters.

A general scheme providing structural materials with self-healing ability for mechanical damages might be as follows:

1. Self-healing agents are added homogeneously or embedded
   by forming microcapsules and tubes into materials.
2. The healing agents are delivered to damage sites
   autonomously under operating conditions of the materials.
3. Damages are healed by reaction between the healing agents
   and materials elements at damage sites under operating
   conditions.

Recently, self-healing of creep cavity and fatigue cavity/crack has been proposed by Shinya [1–7] and Lumley [8, 9] groups, respectively. The self-healing schemes proposed by them for creep cavity and fatigue cavity/crack resemble each other and correspond to the general principle of self-healing for structural materials. Their schemes are as follows:

1. Solute elements with healing functions are added into
   metallic materials and solid solution treated.

Self-healing Materials: Fundamentals, Design Strategies, and Applications. Edited by Swapan Kumar Ghosh
Copyright © 2009 WILEY-VCH Verlag GmbH & Co. KGaA, Weinheim
ISBN: 978-3-527-31829-2

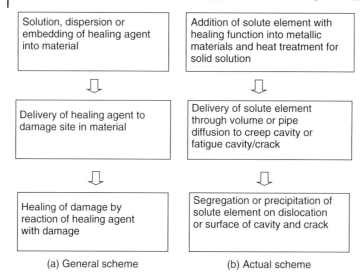

Fig. 7.1 (a) General scheme for self-healing of damage in structural materials and (b) actual scheme for self-healing of creep cavity in heat resisting steel and fatigue cavity/crack in Al alloy.

2. The solute elements are delivered to damage sites autonomously through volume or pipe diffusion during high temperature or fatigue loading services of the materials.

3. The solute elements segregate on dislocations or cavity surfaces or precipitate elemental substances or chemical compounds on surfaces of cavity and crack. The segregated solute elements or the precipitated substances heal the damages by modifying creep cavity surface properties or closing the fatigue cavity/crack.

The general self-healing scheme for mechanical damages in structural materials and the actual self-healing scheme for creep cavity and fatigue cavity/crack are shown in Figure 7.1.

7.2
Self-healing of Creep Cavity in Heat Resisting Steels

Long-term high-temperature services of heat resisting steels lead to premature and low-ductility creep fracture due to cavitation [10]. Creep fracture is caused by nucleation, growth, and coalescence of creep cavities on grain boundaries. It might be possible to prevent most failures in high-temperature structures when the creep cavities are self-healed autonomously during high-temperature service of the steels.

The self-healing method recently proposed by Shinya *et al.* [1–7] is as follows. Creep cavity is thought to grow by diffusive transport of matter from creep cavity surface to grain boundary, which means that the physical property of creep cavity surface is closely connected to cavity growth. Boron (B) segregation and boron nitride (BN) precipitation onto creep cavity surface are thought to improve the physical properties of the creep cavity surface, since B and BN are very stable at high temperatures. It was shown in the works that the segregation of B and precipitation of BN reduced creep cavity growth rate remarkably, which improved long-term creep rupture strength coupled with long-term ductility. The B segregation and BN precipitation onto creep cavity surface are developed autonomously during high-temperature service, suppressing creep cavity growth. It was considered that the segregation and the precipitation provide the steels with function of autonomous self-healing for creep damage.

7.2.1
Creep Fracture Mechanism and Creep Cavity

Fracture mechanisms of metallic materials vary depending on conditions such as temperature and stress. Fracture mechanism maps [11–13], in which this variation in fracture mechanisms and their regions on maps are plotted on stress and temperature axes or stress and time axes, give a bird's-eye view of the variations over a wide range of anticipated conditions of service. Creep fracture of heat resisting steels forms a part of the fracture mechanism map and is often located at boundary regions, where fracture modes or mechanisms vary in complex fashion.

A map of Cr–Mo–V turbine rotor steel is shown in Figure 7.2, which is a typical example of creep fracture mechanism maps for low alloy steels. Three kinds of creep fracture mechanism fields constitute the map.

They are (i) the transgranular creep fracture field located in a relatively short time to rupture region, (ii) the creep cavity-induced intergranular fracture field located in a long time to rupture region, and (iii) the recrystallization rupture field located at higher temperatures. Assuming that the conditions for use of the Cr–Mo–V steel in turbine rotors of power plants are somewhere around 500 °C and 100 MPa, the fracture mechanism of the steel caused in the plant is anticipated, from Figure 7.2, to be creep cavity-induced intergranular creep fracture, which is notable for low ductility and premature rupture life.

Figure 7.3 shows creep cavities observed by a scanning electron microscope (SEM) in specimens interrupted during creep tests at 575 °C and 177 MPa, which is a condition for intergranular creep fracture. Creep cavities generate in a relatively early creep stage at interfaces of precipitates on grain boundaries perpendicular to the stress axis (Figure 7.3a), grow with increasing creeping time along the grain boundaries (Figure 7.3b), and then form intergranular cracks by linking up each others (Figure 7.3c). The intergranular crack propagates further and causes the low-ductility intergranular creep fracture.

Most high-strength heat resisting steels, including austenitic stainless steels, show this type of fracture mechanism—creep cavity-induced transgranular creep

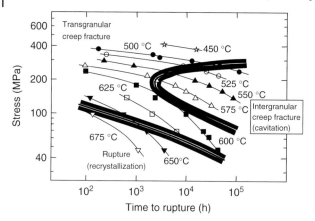

Fig. 7.2 Creep fracture mechanism map for Cr–Mo–V turbine rotor steel, plotted on stress and time to rupture axes with time to rupture curves. Three kinds of creep fracture mechanisms constitute the map: transgranular creep fracture, intergranular creep fracture caused by cavitation, rupture due to dynamic recrystallization.

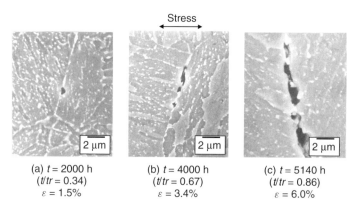

Fig. 7.3 Creep cavities developed in Cr–Mo–V turbine rotor steel during creep test at 575 °C and 177 MPa.

fracture at their actual service conditions. Figure 7.4 shows SEM photographs of developing creep cavities with creep curve during creep test at 750 °C and 37 MPa on a 304 austenitic stainless steel (SUS304H). Creep cavities are formed at interfaces between grain boundary carbides of $M_{23}C_6$ and matrix, and their sizes at the first stage of creep rupture life ($t/tr = 0.3$) are less than 0.5 μm. These fine creep cavities grow along grain boundary perpendicular to tensile stress, leading to lens- or rod-like shapes at the second stage ($t/tr = 0.5$). Further growth and linking up each others lead to grain boundary cracks of about 5 μm at the third

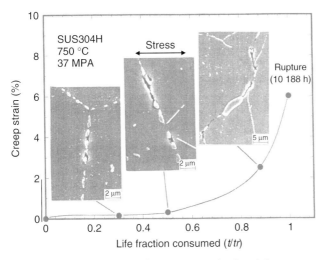

Fig. 7.4 Typical appearances of creep cavities developed during creep exposure with creep curve, observed in austenitic stainless steel. Creep cavities nucleate at $M_{23}C_6$ carbides on grain boundaries, grow along grain boundary, and form grain boundary cracks due to linking up each others.

stage ($t/tr = 0.9$). From Figures 7.3 and 7.4, it is indicated that size of creep cavities at early stages of their developments are very fine, which suggests that application of healing methods may be effective at the early stages.

7.2.2
Sintering of Creep Cavity at Service Temperature

Fine creep cavities at initial stages may be possible to be removed through the sintering treatment of short-time holding at usual service temperatures under reduced or removed stress loading in high-temperature plants. It is estimated that grain boundary diffusion is sufficiently active for sintering the fine creep cavities on grain boundaries at the usual service temperatures [14, 15]. Provided the short-time holding at service temperatures removes the creep cavities, it is possible to extend the service lives of high-temperature plants and structures significantly. Hence, trial experiments were carried out to clarify the possibility of removing creep cavities by sintering at service temperatures.

In order to estimate sintering rates of creep cavities at operating temperatures of heat resisting steels, damaged specimens with creep cavities were prepared. Specimens of a 1.3Mn–0.5Mo–0.5Ni steel for boiler use [15] were crept at 550 °C and 118 MPa for 540 h ($t/tr = 0.6$) to introduce creep cavities at grain boundaries. The mean size and spacing of the introduced creep cavities were 1.8 and 2 μm, respectively. These values were used for theoretical sintering calculations. Total amount of creep cavity volume was determined by measuring change in density [16]. A change in density ($\Delta D/D$) of the damaged sample used for the study was -0.56×10^{-3}. The sectioned samples with creep cavities were annealed up to 800 h, at 550, 600, and 700 °C, respectively. After annealing, the ratios of sintered creep cavities to initial ones were evaluated by change in density recovered from the initial value (-0.56×10^{-3}).

For calculation of sintering rate of creep cavities, the theory of grain boundary diffusion controlled creep cavity growth by Speight and Beeré [17] was adopted. The following equation, which gives shrinking rate of creep cavity, is derived by substituting zero for the applied tensile stress ($\sigma = 0$) in the original creep cavity growing equation by Speight and Beeré [17]:

$$\frac{dr}{dt} = \frac{\Omega D_{gb} \delta \gamma_s}{r^3 kT \left[\dfrac{\ln \lambda}{2r} - \dfrac{1}{4} \left(1 - \dfrac{4r^2}{\lambda^2} \right) \left(3 - \dfrac{4r^2}{\lambda^2} \right) \right]} \tag{7.1}$$

where t is the time, r is the half of creep cavity diameter, λ is the creep cavity spacing, δ is the grain boundary thickness, T is the absolute temperature, Ω is the atomic volume, γ_s is the surface free energy, D_{gb} is the grain boundary diffusion coefficient, and k is the Boltzmann constant.

The creep cavities are assumed to be uniform size spheres and distributed evenly on grain boundaries perpendicular to tensile stress, and their determined initial diameter and spacing are 1.8 and 2 μm, respectively. Figure 7.5 [15] shows the comparison between calculated sintering curves based on grain boundary diffusion controlled model and experimental annealing data. It shows that the experimental sintering rate is $1/10^4$ times as low as the calculated value. The calculated sintering rate is sufficiently rapid for removing the creep cavities through annealing treatment at usual service temperatures of the steels, whereas actual sintering rate with experimental annealing is too slow to be actually used for removing creep cavities.

When compressive stress or hydrostatic pressure is applied during experimental sintering annealing, the sintering rate is accelerated drastically [14, 15]. The sintering rate during simple annealing was thought to have been delayed by suppression of the matter flow from grain boundary to creep cavity, which is caused by tensile stress generated due to the material flow as shown in Figure 7.6. By the tensile stress acting on grain boundary, the flow of matter is constrained, and further sintering is suppressed. The loading compressive stress removes the tensile stress, which lets further sintering to continue. It is difficult to effectively sinter the creep cavities only by holding at usual operating temperatures for short while under reduced or removed operating stress loading. Sintering at higher temperatures, where volume diffusion is active, or under compressive stresses might remove the creep cavities more effectively. These sintering treatments, however, are not realistic to put into practice in high-temperature plants and structures.

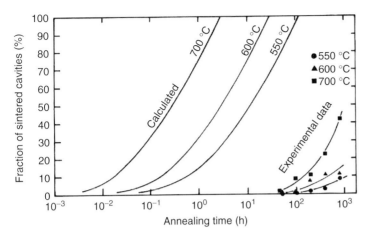

Fig. 7.5 Comparison between calculated sintering rates based on grain boundary diffusion model and experimental data of annealing for creep cavities induced in 1.3Mn–0.5Mo–0.5Ni steel. Mean size and spacing of initial creep cavities are 1.8 μm and 2.0 μm, respectively.

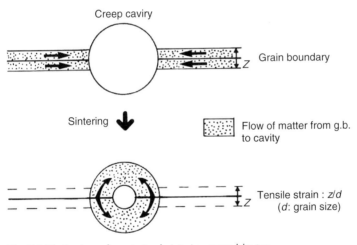

Fig. 7.6 Mechanism of constrained sintering caused by tensile strain at grain boundary associated with flow of matter from grain boundary to surface of creep cavity.

7.2.3
Self-healing Mechanism of Creep Cavity

7.2.3.1 Creep Cavity Growth Mechanism

The creep rupture life and ductility in the region of cavitation field fracture depend on the development of creep cavity [12]. Although the development of creep cavity depends on nucleation and growth rates of creep cavity, the creep cavity growth

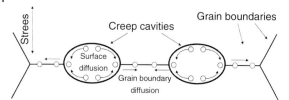

Fig. 7.7 Illustration of creep cavity growth process controlled by grain boundary or surface diffusion wherein atomic transport occurs along the creep cavity surface and then down the grain boundary. Slower diffusion rate controls creep cavity growth.

is thought to be more influential since most creep cavity nucleation is believed to occur at initial stages of creep rupture life [18].

Figure 7.7 shows an illustration of creep cavity growth mechanism, which is thought to act in the usual cavitation range, especially at lower applied stresses and higher temperatures. The creep cavity growth proceeds with diffusive transport of matter from creep cavity surface onto grain boundary, where they can be deposited. The deposition of matter on the grain boundary causes reduction of tensile stress at grain boundary, which is a driving force for the flow of matter. In usual cavitation region, the stress acting on grain boundary in association with the deposition is compensated by tensile creep deformation and diffusion of the deposited matter through grain boundary, which makes it possible to continue unconstrained growth of creep cavity. Needham and Gladman [19] extensively studied the creep cavitation behaviors of a type 347 austenitic stainless steel. They concluded that the measured creep cavity growth rate could be well predicted by the unconstrained diffusive growth mechanism, considering the creep strain dependence of continuous creep cavity growth nucleation. Under unconstrained conditions, diffusivities along both grain boundary and creep cavity surface control creep cavity growth rate.

From Figure 7.7, the creep cavity growth rate is expected to be controlled by the slower process of either grain boundary diffusion or creep cavity surface diffusion [20]. The creep cavity growth under unconstrained condition is expected to be substantially influenced due to the change in self-diffusion coefficient along both creep cavity surface and grain boundary, which are significantly influenced by segregation of trace elements. It may be possible to reduce the surface diffusion rate by modification of creep cavity surface, leading to reduction of creep cavity growth rate. Hence, the modification of creep cavity surface was selected as a focusing target for a trial study of self-healing of creep cavity.

7.2.3.2 Self-healing Layer on Creep Cavity Surface

The creep cavity surface diffusion is known to be influenced by segregation of trace elements on its surface [21]. Some trace elements diffuse to grain boundary and also to creep cavity surface, and segregate there at high temperatures. It is also known that soluble S segregates onto creep cavity surface very easily, which gives rise to accelerated creep cavity growth through increase in surface diffusion rate [21]. Owing to the low melting point of S (112.8 °C), the creep cavity surface contaminated with S becomes very active and surface diffusion rate increases by several orders of magnitude [21–23].

When the soluble S is almost completely removed by addition of Ce and Ti [24, 25], other elements such as B and N become possible to segregate onto creep cavity surface. Segregation of B is thought to suppress the surface diffusion at creep cavity surface, reducing the creep cavity growth rate, owing to the high melting point of B (2080 °C). Cosegregation of B and N might form BN compound at creep cavity surface. The BN precipitation on creep cavity surface is expected to suppress the surface diffusion almost completely and then retard the creep cavity growth significantly owing to its higher melting point (3000 °C).

In usual heat resisting steels, a trace of S segregates onto creep cavity surface, and accelerates the surface diffusion rate of creep cavity significantly. Since the accelerated surface diffusion rate is by far faster than the grain boundary diffusion rate, the creep cavity growth rate is controlled by the grain boundary diffusion rate in the usual steels. The surface layers of segregated B and precipitated BN is thought to significantly reduce the surface diffusion of creep cavity. The layer of BN precipitation, particularly, is thought to almost freeze the diffusion rate. From the significant reduction in the diffusion rate, the surface layers should change the controlling diffusion process for the creep cavity growth from the grain boundary to the surface diffusion. The creep cavity growth rate controlled by the surface diffusion [26, 27] is estimated as:

$$\frac{dr}{dt} = \Omega D_s \delta \gamma_s \left\{ \sigma_b / [(1 - \omega)\gamma_s \sin(\psi/2)] \right\}^3 / kT \tag{7.2}$$

where D_s is the surface diffusion coefficient, σ_b is the normal stress acting on the grain boundary, $\omega = (2r/\lambda)^2$, and ψ is the angle at the chip of creep cavity.

Both the B segregation and the BN precipitation on creep cavity surface occur continuously and cover entire creep cavity surface autonomously during usual high-temperature service of the heat resisting steels. These surface layers of B and BN are self-healed by continuous segregation or precipitation during the service. In addition to the self-healing of themselves, the surface layers provide the steels with self-healing function for cavitation since they autonomously cover the creep cavity surface and suppress the cavitation almost completely. Figure 7.8 shows the formations of the self-healing layers for creep cavity by B segregation and BN precipitation.

7.2.4
Self-healing of Creep Cavity by B Segregation

7.2.4.1 Segregation of Trace Elements
Effects of trace elements on creep rupture properties, studied up to the present time, are summarized as follows. Holt and Wallace [28] have classified the most common trace elements according to whether they have detrimental or beneficial effect on creep rupture strength. Among them, S and O can cause the most severe

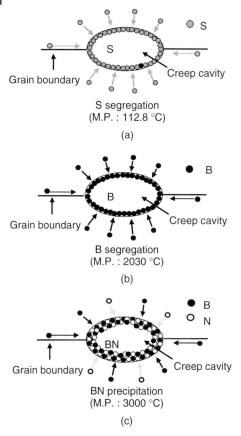

Fig. 7.8 Illustration of self-healing layer formation of B segregation and BN precipitation on creep cavity surface. In usual steels, trace of S segregates onto creep cavity surface during high-temperature service. In self-healing heat resisting steels, the layers of segregated B and precipitated BN suppress creep cavity growth.

embrittlement during creep, even at low parts per million levels. These elements in solid solution, even in minute quantities, must be controlled, either by removing them during melting the steel or by alloying the steel with suitable elements to precipitate them out, so as to increase creep rupture strength and ductility. Minor addition of rare earth elements, such as Ce, is highly effective in removing S and O in steel through formation of ceriumoxysulfide (Ce_2O_2S) [24].

Boron addition in high-temperature alloys has been reported to increase creep rupture strength. It is widely believed that the addition of B increases creep rupture life and ductility through the increase in creep cavitation resistance of the steels by grain boundary strengthening. The reason for strengthening of the grain boundary is not precisely known. In most instances, it is thought that B is concentrated on grain boundary, where it enters into the precipitates or matrix/precipitates on the grain boundary in such a way to suppress microcavity formation. Although many beneficial effects of B have been reported, actual effects and behaviors of B are not certain, particularly on cavitation and creep fracture, due to their complexity.

Effects of trace elements on grain boundary behaviors have been extensively studied, whereas the knowledge concerned with cavitation is limited. In the case of creep fracture mechanism and property, the physical state of creep cavity surface is thought to influence more directly and effectively than that of grain boundary. Modification of creep cavity surface, which might be very effective on suppression of creep cavity growth, has been tried by segregation of B and precipitation of BN.

7.2.4.2 Self-healing of Creep Cavity by B Segregation onto Creep Cavity Surface

In order to facilitate B to segregate onto creep cavity surface, a chemical composition of a 347 austenitic stainless steel was modified, which was melted in a vacuum arc furnace. Using the modified austenitic stainless steel, the presence of B segregated onto creep cavity surface and the self-healing effect of B segregation on creep cavity growth were proved. Table 7.1 shows chemical compositions of a standard 347 and the modified 347 austenitic stainless steels. In the modified 347 steel (347BCe), the content of free S was almost completely removed by the addition of 0.016 mass% of Ce, which has a strong affinity to S and O, leading to formation of Ce_2O_2S [24] and Ce_2S_3. In addition to Ce, 0.07 mass% of B was also added to the modified 347 steel. The steels were given a solution heat treatment at 1200 °C for 20 min, followed by water quenching.

Creep rupture tests were carried out at 750 °C. The variation of creep rupture life and ductility of both the steels with applied stress are shown in Figure 7.9. Addition of minute amount of B and Ce had remarkable effects on the creep rupture strength and ductility of the modified 347 steel. Creep rupture strength and ductility of the steel increased with the addition of B and Ce, effects of which were pronounced on longer creep exposure.

Figure 7.10 shows creep cavities observed in the ruptured specimens of the standard 347 and the modified 347 (347BCe) steels. Creep cavities in the standard 347 steel grow quickly along grain boundary and form grain boundary cracks by linking up each others, whereas creep cavities in the modified 347 steel remain fine size and isolated from each others.

X-ray diffraction analyses of the precipitates were carried out in both the steels after rupture at 750 °C and 69 MPa. Precipitate residues were extracted from the steels by the electrochemical method. Figure 7.11 indicates the presence of

Table 7.1 Chemical compositions of standard 347 (B-free steel) and modified 347 (B-added steel) austenitic stainless steels (mass %)

Alloy	C	Si	Mn	P	S	Cr	Ni	Nb	N	B	Ce
Standard 347	0.080	0.59	1.68	0.001	0.002	17.96	12.04	0.41	0.077	—	—
Modified 347	0.078	0.68	1.67	0.001	0.002	18.15	11.90	0.38	0.072	0.069	0.016

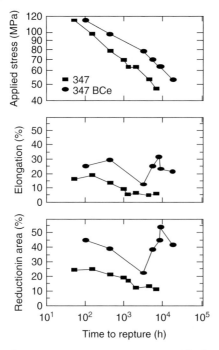

Fig. 7.9 Creep rupture properties at 750 °C for standard 347 and modified 347 (347BCe) austenitic stainless steels. Modified 347 steel shows higher rupture strength and ductility, particularly for longer rupture life.

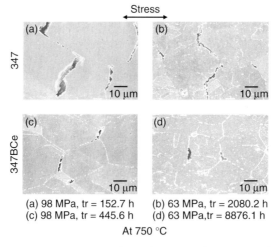

(a) 98 MPa, tr = 152.7 h   (b) 63 MPa, tr = 2080.2 h
(c) 98 MPa, tr = 445.6 h   (d) 63 MPa,tr = 8876.1 h

At 750 °C

Fig. 7.10 Creep cavities observed in ruptured specimens of standard 347 and modified 347 (347BCe) austenitic stainless steels.

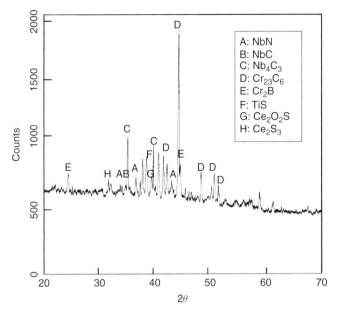

Fig. 7.11 X-ray diffraction analysis of the precipitates extracted by electrochemical method from the B-containing modified 347 steel, creep tested at 750 °C and 69 MPa.

different precipitates in the B-containing modified 347 steel. In both the steels, the presence of $Cr_{23}C_6$, Nb_4C_3, NbC, and NbN precipitates were observed. The observed carbonitride precipitation in both the steels was reported by several investigators on the type 347 austenitic stainless steel [29, 30]. X-ray investigations also indicated the presence of Ce_2S_3 and Ce_2O_2S precipitates in the modified 347 steel. The addition of Ce in steels may be an effective way of removing traces of S and O soluble in the steels. In addition to Ce_2S_3 and Ce_2O_2S, it was observed that a trace of Ti forms TiS, showing that Ti also has a strong affinity to S. Chromium boride (Cr_2B) particles were observed in the B-containing steel. The solid solubility of B in a 18 Cr–15 Ni austenitic stainless steel has been reported to be about 90 ppm at 1000 °C and the solubility boundary recedes rapidly with decreasing temperature, which indicates high stability of the precipitates in the steel.

Interrupted creep tests at 78 MPa and 750 °C were carried out in argon atmosphere to measure the growth rate of creep cavities observed on the specimen surface. Individual creep cavity dimensions had been measured by SEM on interruption of the creep tests and the average growth rates of the creep cavity for the intervals were calculated. Measurements were carried out on several creep cavities until they coalesce with each others. Figure 7.12 compares the creep cavity growth rate of the modified steel with that of the standard steel until they coalesce with each others. Addition of B in the modified steel decreased the creep cavity growth rate almost by an order of magnitude.

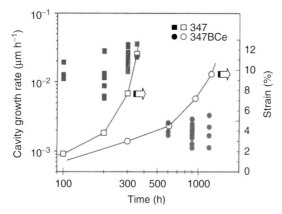

Fig. 7.12 Creep cavity growth rate and creep strain with creep exposure time in Ar at 750 °C and 78 MPa. The growth rate in modified 347 steel is slower by an order of magnitude than that of standard 347 steel.

The chemistry of the creep cavity surface in both the steels was examined by an Auger electron spectroscope (AES). The AES is sensitive only to the top few atom layers on fractured surface, making it a useful technique for studying any trace element segregation that might have occurred. The crept specimens of the steels were fractured by impact loading at liquid nitrogen temperature in the AES chamber to expose the creep cavity surface. Figure 7.13 shows fractured surfaces with creep cavities on grain boundaries and Auger spectra obtained from the creep cavity surfaces of the standard and modified 347 steels. The fractured surface of both the steels containing creep cavities on grain boundary indicates that this procedure is effective in exposing the creep cavity surface.

Presence of S segregation was observed on the creep cavity surface of the standard 347 steel, whereas Auger peak of elemental B instead of S was observed in the modified 347 steel. The soluble S in the modified 347 steel might be removed almost completely with the addition of Ce, by forming $Ce_2O_2S_2$ and Ce_2S_3. In the absence of S contamination, most nucleated creep cavity surfaces were covered with filmy layer of segregated elemental B. Figure 7.14 shows Auger S and B mapping of the fractured surfaces. It is indicated that S segregates onto creep cavity surfaces and covers up the surfaces in the standard 347 steel, whereas in the B-containing modified 347 steel, B segregates onto the creep cavity surfaces. The S content in the standard 347 steel was ∼0.002 mass% (Table 7.1) and even such amount of bulk S could contaminate the creep cavity surface. White *et al.* [21], in their study on effects of S and P on creep cavitation in a 304 austenitic stainless steel, reported S concentration on the creep cavity surface approximately 103 times than that in the bulk. The strong tendency of B to segregate on interfaces is derived from an effect of large misfit of B atom in both substitutional and interstitial sites of the austenitic lattice.

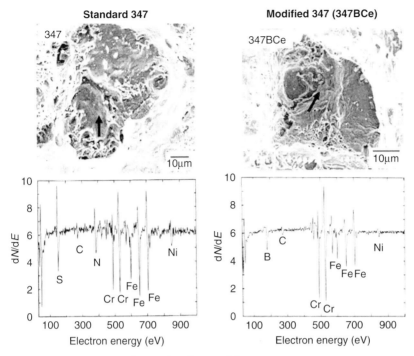

Fig. 7.13 Creep cavities on fractured surfaces and Auger spectra from creep cavity surfaces of standard 347 and modified 347 steels creep exposed at 750 °C and 69 MPa. The crept specimens were fractured by impact loading at liquid nitrogen temperature in AES chamber.

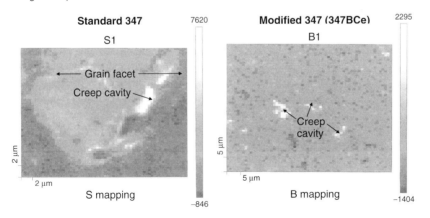

Fig. 7.14 Auger S and B mapping of the fractured surfaces, creep exposed at 750 °C and 69 MPa. In standard 347 steel, extensive segregation of S is shown on creep cavity surface and slight segregation on grain facet, and in modified 347 steel segregation of B is noticed only on creep cavity surface.

7.2.4.3 Effect of B Segregation on Creep Rupture Properties

The B segregation is expected to decrease the diffusivity along the creep cavity surface due to its relatively high melting point of around 2080 °C. The creep cavitation in the modified 347 steel (Figures 7.10 and 7.12) is thought to be suppressed by the segregation of B on the creep cavity surface. The suppression should prevent from the grain boundary fracture and increase significantly the creep rupture strength associated with high ductility.

The segregation of B on creep cavity surface occurs autonomously and covers up the surface during high-temperature service of the steel. The surface filmy layer of the B segregation prevents from creep fracture by suppressing creep cavity growth and provides the steels with superior creep rupture properties.

7.2.5
Self-healing of Creep Cavity by BN Precipitation on to Creep Cavity Surface

7.2.5.1 Precipitation of BN on Outer Free Surface by Heating in Vacuum

The compound of BN is known to precipitate on outer free surface of B-containing austenitic stainless steels by heating in vacuum. Stulen *et al.* [31] studied segregation of B onto outer free surface using an austenitic stainless steel of 21Cr–6Ni–9Mn, containing a bulk concentration of less than 10 ppm of B. The segregation results with heating in vacuum show that B exhibits a strong surface activity at temperatures above 700 °C. Above 700 °C, B and N cosegregate to form a surface layer of BN, which extends into the bulk of the steel. The results on 304L and 304LN further illustrate that the high concentration enhancement for B at the surface is due to the strong attractive B–N interaction. Thus, the surface N functions essentially as a chemical trap for the B, tying it to the surface. Nii *et al.* [32, 33] also observed the precipitation of BN by heating in vacuum on outer free surface of a 304 austenitic stainless steel doped with N, B, and Ce. They confirmed that the surface layer of BN uniformly covers almost the entire outer free surface by means of heating at 730 °C in vacuum, and also the layer itself is self-healed by heating in vacuum again.

It may be possible to precipitate the BN thin layer on creep cavity surface during high-temperature service of B-containing austenitic stainless steels, since the inner surface of creep cavity is almost a vacuum. The precipitated BN thin film on creep cavity surface is expected to reduce creep cavity growth rate through suppressing surface diffusion rate at creep cavity surface since melting point of BN is considerably high (3000 °C).

7.2.5.2 Self-healing of Creep Cavity by BN Precipitation

A 304 austenitic stainless steel was modified with the additions of 0.07 mass% of B, 0.064 mass% of N, 0.33 mass% of Ti and 0.008 mass% of Ce, aiming at precipitation of BN onto creep cavity surface. Chemical compositions of the melted standard 304 and the modified 304 (corresponding to type 321) austenitic stainless steels are shown in Table 7.2. The additions of Ce and Ti are expected to remove soluble S through the formation of Ce_2O_2S [24] and $Ti_4C_2S_2$ [25], respectively. In the absence of S segregation, B and N segregate onto the creep cavity surface simultaneously

Table 7.2 Chemical compositions of standard 304 and modified 304 austenitic stainless steels (mass %).

Steel	C	Si	Mn	P	S	Ni	Cr	B	N	Ti	Ce
Standard 304	0.082	0.49	1.62	0.021	0.009	10.05	19.07	—	0.0072	—	—
Modified 304	0.096	0.50	1.52	0.020	0.002	10.07	19.16	0.070	0.0635	0.33	0.008

and form BN compound there. The standard 304 steel was subjected to a solution heat treatment at 1130 °C for 29 min, whereas the modified 304 steel to a solution heat treatment of 1180 °C for 20 min.

In preliminary experiments of BN precipitation, the B-containing modified 304 steel was tensile tested, and a specimen including cavities in a necked zone was obtained by stopping the test just after forming a necking zone and before fracture. Samples taken from the specimens were heated for 120 min at 750 °C, and then fractured under impact loading at liquid nitrogen temperature in AES chamber. On the fractured surface, tensile cavities were observed. From the exposed tensile cavity surface, Auger spectra were obtained. The tensile cavities on the fractured surface and the obtained Auger spectrum are shown in Figures 7.15 and 7.16. The sharp and high peaks of B and N indicate that B and N segregate simultaneously and cover extensively the tensile cavity surface. From the positions and shapes of B and N peaks, which are the same as those reported by Stulen *et al.* [31], it was concluded that the cosegregated B and N form the stable compound of BN [31, 33]. The surface of tensile cavity should be covered extensively with the BN layer since peaks of other elements were not observed (Figure 7.16). The present preliminary experiment suggested that the stable compound of BN should precipitate onto creep cavity surface during creep exposure of the steel.

The results of creep rupture tests carried out at 750 °C are shown in Figure 7.17. It indicated that the addition of minute amount of B and Ce significantly increases creep rupture strength and ductility, which is more pronounced on longer creep exposure.

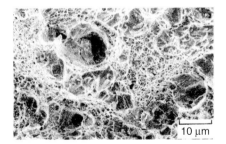

Fig. 7.15 Tensile cavities observed by SEM on fractured surface of tensile tested specimen of B-containing modified 304 steel.

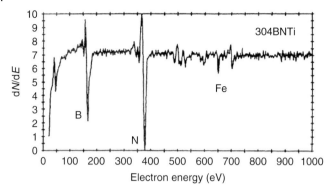

Fig. 7.16 Auger spectrum obtained from surface of tensile cavity in tensile tested specimen heated for 120 min at 750 °C. The positions and shapes of B and N peaks indicate formation of BN compound on the surface.

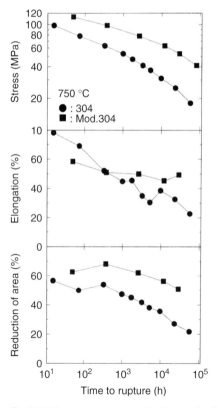

Fig. 7.17 Creep rupture properties (strength, elongation, and reduction of area) at 750 °C of standard 347 and modified 304 steels. Modified 304 steel shows higher rupture strength and ductility, particularly for longer rupture life.

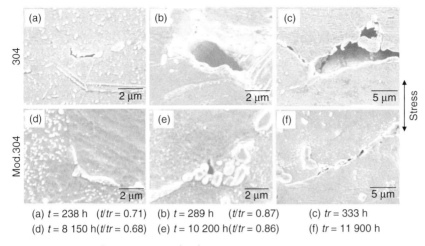

(a) $t = 238$ h ($t/tr = 0.71$) (b) $t = 289$ h ($t/tr = 0.87$) (c) $tr = 333$ h
(d) $t = 8\ 150$ h($t/tr = 0.68$) (e) $t = 10\ 200$ h($t/tr = 0.86$) (f) $tr = 11\ 900$ h

Fig. 7.18 Scanning electron micrographs showing creep cavities in specimens crept at 750 °C and 63 MPa. Creep cavity growth in modified 304 steel is remarkably suppressed.

The creep cavities in the ruptured specimens of the modified 304 steel are few and fine as shown in Figure 7.18 when compared to those of the standard 304 steel. It is indicated that the creep cavity growth is remarkably suppressed in the modified 304 steel. The suppression should give the improved creep rupture properties due to preventing the grain boundary fracture. The crept specimens were fractured by impact loading at liquid nitrogen temperature in the AES chamber to expose the creep cavity surface.

Figures 7.19 and 7.20 show facture surfaces with creep cavities and Auger spectra obtained from the creep cavity surfaces in the standard and modified 304 steels, respectively. The sharp and high peak of S was observed in the standard 304 steel. In the modified 304 steel, the peaks of B and N on the creep cavity surface were observed, whereas the presence of S was not observed. The energy positions and the shapes of B and N peaks indicate that the segregated B and N form the stable compound of BN [31, 33]. In addition to the peaks of B and N, comparatively high peaks of C and O, resulting from the contamination, were observed. To find the suitable creep cavities for the AES observation on the fractured surface, it took longer time, since the creep cavities are few and fine in the steel. The delay might be the reason for the appearance of higher peaks of C and O, which are main contaminates in the residual gas in the AES chamber.

The creep cavities in the notched specimen of the modified 304 steel were subjected to significant deformation by impact loading. The deformation may also be the reason why the peaks of B and N are comparatively low when compared with that of tensile cavity (Figure 7.16). Figure 7.21 shows the shapes of elemental B peak in the modified 347 steel and B and N peaks of BN in the modified 304 steel. The difference in Auger peaks of elemental B and B as BN is shown clearly.

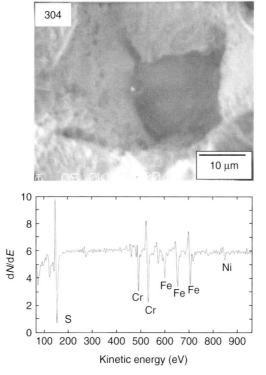

Fig. 7.19 Scanning electron micrograph showing creep cavity surface, exposed by breaking at liquid nitrogen temperature under impact loading, of standard 304 steel crept for 290 h ($t/tr = 0.87$) at 750 °C and 63 MPa and Auger spectrum obtained from the creep cavity surface.

The X-ray diffraction patterns were obtained from the precipitate of the modified 304 steel after creep rupture tests.

The pattern shown in Figure 7.22 indicates the presence of $Ti_4C_2S_2$ and Ce_2O_2S. The coaddition of Ce and Ti should be highly effective in removing the soluble S in the steel and in preventing from depositing S on the creep cavity surface. In the absence of S contamination, B and N segregate onto creep cavity surface and form the filmy BN precipitates, which significantly reduce the creep cavity growth rate (Figure 7.18). The suppression of the creep cavity growth rate provides the modified 304 steel with higher rupture strength and ductility, particularly for longer rupture time region (Figure 7.17).

7.2.5.3 Effect of BN Precipitation on Creep Rupture Properties
Comparison of creep rupture properties of the present modified 304 austenitic stainless steel and conventional 304 and 321 austenitic stainless steels for high-temperature use are shown in Figure 7.23. The plots for the conventional 304 and 321 austenitic stainless steels include data on nine steels for each of

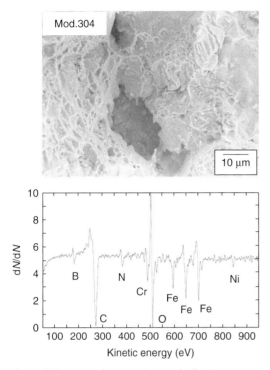

Fig. 7.20 Scanning electron micrograph showing creep cavity surface, exposed by breaking at liquid nitrogen temperature under impact loading, of modified 304 steel crept for 10 200 h (t/tr = 0.86) at 750 °C and 63 MPa and Auger spectrum obtained from the creep cavity surface.

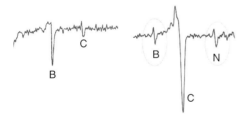

Fig. 7.21 Comparison of Auger peak shapes between segregated B in modified 347 steel and segregated B and N in modified 304 steel. Different peak shapes of elemental B and B as BN compound are shown.

the two types of steels, respectively, which were creep rupture tested at National Institute for Materials Science, Japan. Absolute higher creep rupture strength and ductility of the modified 304 steel (304BNTi) than that of SUS 304 H (18Cr–8Ni) and SUS 321 H (18Cr–10Ni–Ti) steels were indicated, particularly for prolonged creep rupture life region. This remarkable improvement in rupture properties is

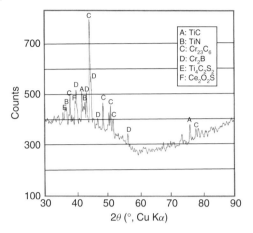

Fig. 7.22 X-ray diffraction analysis of precipitate, extracted by electrochemical method from B-containing modified 304 steel, creep tested at 750° C and 63 MPa.

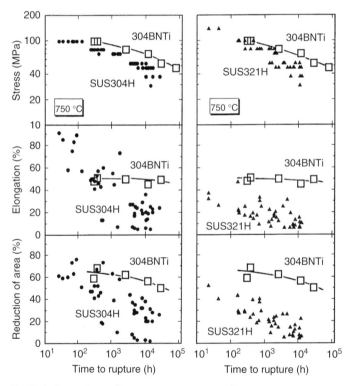

Fig. 7.23 Comparison of creep rupture properties between modified 304 steel (304BNTi) and conventional 18-8-based austenitic stainless steels (304 and 321).

thought to be derived from the precipitation of BN onto creep cavity surface. It is clearly indicated that the BN precipitation is an excellent method not only to extend lives of high-temperature structures by providing them with self-healing ability but also develop superior heat resisting steels by preventing the notorious grain boundary creep fracture. The comparison of creep rupture properties between the modified 347 steel (Figure 7.9) and the modified 304 steel (Figure 7.17) indicates that the precipitation of BN gives rise to the higher rupture strength and ductility, particularly at longer rupture life, than that of B segregation. The higher melting point and finer layer of BN than that of elemental B should be the reason for the superiority.

7.3
Self-healing of Fatigue Damage

Accumulated fatigue damage causes failures of bridges, airplanes, cars, and trains, and in some cases leads to serious disasters affecting people's lives. Fatigue crack, however, is not easy to be detected in its initial stage owing to its fine size, and also very difficult to be repaired on service sites of the structures. This is the reason why especially self-healing of fatigue crack is expected to be developed and put into practical use for prevention of failures and the life extension of the structures.

A high hurdle for self-healing of fatigue crack is how healing agents should be delivered to crack sites, particularly in metallic materials, since volume or grain boundary diffusions are not active at ambient temperature of structures. Recently, Lumley et al. [8, 9] have found out that solute Cu atoms in Al alloys are mobile and able to arrive at crack sites by pipe diffusion through dislocation cores at room temperature [34]. The delivered solute Cu atoms should precipitate at crack surfaces and close the cracks. Lumley's recent works on self-healing of fatigue crack are introduced here together with related reports.

7.3.1
Fatigue Damage Leading to Fracture

Metallographic observations of the fatigue process in metallic materials have shown that much of the associated deformation with fatigue loading is not homogeneous and tend to be concentrated in localized regions in the material. Usually, a fatigue crack initiates from the deformation concentration region, and is observed at the sites of (i) fatigue slip bands, (ii) grain boundaries, and (iii) inclusions. Common for the three types of nucleation is local plastic strain concentration at or near the surface [35]. It is thought that nucleation in the fatigue slip bands is a basic type of nucleation not only because this is the most frequent case but also mainly because the cyclic slip process and formation of fatigue slip bands may also precede nucleation of crack at grain boundaries and inclusions.

The concentration of deformation in certain slip planes forms intense lines and occasionally wider striations on the planes. The intense slip lines and slip striations frequently contain fine cavities and cracks [36]. It is commonly observed that an arrayed line of discrete cavities is present ahead of an advancing crack, suggesting that crack growth occurs by joining of the cavities. This is the reason why the fatigue crack is called *fatigue cavity/crack* in this chapter. It is thought that a large number of vacancies are generated in slip bands and slip striations, and that cavities may be formed by condensation of some of these vacancies.

Taking in to consideration the morphology and the formation process, it may be possible to heal the fine fatigue cavities at initial stages before growing into fatigue cracks by application of the pipe diffusion in similar way as the healing of creep cavities in heat resisting steels.

7.3.2
Delivery of Solute Atom to Damage Site

In self-healing of fatigue damage in metallic materials, the delivery process of solute atom to damage site is the most difficult to accomplish and forms an insuperable barrier (Figure 7.1). Recently, it has been found that the delivery of solute atom to damage site may be possible at ambient temperature through pipe diffusion and solute-vacancy complex formation.

7.3.2.1 Pipe Diffusion

Pipe diffusion along dislocation is known to be much faster than both volume and grain boundary diffusion at comparatively lower temperatures. Pipe diffusion of solute atom through dislocation core is typically 10^5 to 10^6 times faster than volume diffusion in the bulk material at ambient temperature [34], and calculation of solute Cu atom in Al alloy estimates a similar increase of up to 10^6 times at the temperature [37]. This value indicates that pipe diffusion rate of solute Cu atoms in Al alloys corresponds to volume diffusion rate at service temperatures of resisting steels. Such high solute mobility of solute Cu atoms segregated to dislocations suggests that significant quantities of the atoms have the ability to be readily delivered to fatigue cavity/crack sites and contribute to closing the cavity/crack by precipitating at the sites.

Hautakangas *et al.* [38] quantified the amount of the restoration of free volume defects (vacancies, dislocations, and nano cracks) with positron lifetime spectroscope, and from the results concluded that solute Cu atoms in Al alloys are capable of closing the fatigue cavity/crack. Plastic deformation of an underaged Al alloy saturated with solute Cu and Mg atoms increases the average positron lifetime, and after deformation the positron lifetime value decreases with time and reaches its initial value at a rate depending on aging temperature. The temperature dependence of the positron lifetime values suggests that the decay rate is controlled by pipe diffusion of Cu or Mg atoms along dislocation core. They indicated that the restoration of the deformation damage (vacancies, dislocations, and nano cracks) takes place via pipe diffusion of solute Cu or Mg atoms.

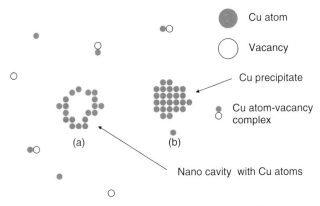

Fig. 7.24 Cu-vacancy complexes: (a) nano cavity covered with Cu atoms and (b) Cu precipitate formed by annealing after irradiation [39, 40].

7.3.2.2 Solute-vacancy Complexes

In addition to the pipe diffusion, solute atoms may be possible to move in metallic materials and be delivered to damage sites at comparatively lower temperatures by forming solute atom-vacancy complexes.

Nagai *et al.* [39–42] obtained samples of Fe-0.3 mass% Cu and Fe-0.05 mass% Cu irradiated at $100\,^{\circ}$C by fast neutrons, and studied the behavior of irradiation-induced vacancies and Cu aggregations with positron lifetime spectroscopy. The results indicate that irradiation-induced vacancies move freely in Fe–Cu materials and bound to solute Cu atoms. The solute Cu atom-vacancy complexes are also mobile and aggregate to nano cavity during the irradiation. Figure 7.24 shows the formations of nano cavity and Cu precipitate, which are typical aggregations of the complexes. The solute Cu atoms segregate on the surface of the nano cavities and cover inside of the nano cavity, since surface energy of Cu is lower than that of Fe. Ultrafine Cu precipitates grow on the nano cavities by annealing after the irradiation. It is well known that the aggregation and Cu precipitation cause embrittlement of nuclear reactor pressure vessel steels.

Their findings on the solute Cu atom behavior may give some hints on the delivery of solute atoms to damage sites and the closure of fatigue cavity/crack.

7.3.3
Self-healing Mechanism for Fatigue Cavity/Crack

Lumley *et al.* [8, 9] proposed the following three microstructural approaches for self-healing of fatigue cavity/crack in a solute element containing underaged Al alloy:

1. Fatigue cavity/crack closure due to heterogeneous precipitation onto the surface in a manner analogous to precipitation-induced sintering.

2. Fatigue cavity/crack closure due to volume expansion causing localized compression, associated with heavy precipitation in plastic zone of cavity/crack.
3. Continual replenishment of strengthening phases within microstructures by dynamic precipitation.

The above three mechanisms—precipitation onto fatigue cavity/crack surface, volume expansion associated with precipitation and replenishment of strengthening phases by dynamic precipitation—are fundamental processes for self-healing of fatigue cavity/crack.

7.3.3.1 Closure of Fatigue Cavity/Crack by Deposition of Precipitate

The contribution of precipitation to porosity closure was studied in Al powder alloys (Al–8Zn–2.5 Mg–1Cu) [43]. The porosity closure was caused by precipitation of the η phase ($MgZn_2$), particularly on pore surfaces. In a similar way to the porosity closure, it may be possible to close the fatigue cavity/crack due to precipitation of solute atoms, which are delivered to the cavity/crack surface through pipe diffusion, as shown in Figure 7.25. The repeated deposition of the precipitate on the surface of fatigue cavity/crack could fill the inside of cavity/crack.

7.3.3.2 Closure of Fatigue Cavity/Crack by Volume Expansion with Precipitation

Large volume expansion is observed in Cu-containing Al alloys [44] with precipitation of θ' phase ($CuAl_2$), and the volume expansion increases proportionally to the amount of θ' phase. This suggests that concentrated dynamic precipitation of θ' phase on dislocations in fatigue slip bands will also cause a localized heterogeneous expansion around the bands. With the concentrated precipitation, tensile stress around the fatigue cavity/crack is relaxed and moreover compressive stress will be generated, which may slow the growing rate of fatigue damage [9].

Dimensional change associated with precipitation of θ' phase was measured with differential dilatometer [9], which showed the expansion from -100 to $1300 \, \mu mm^{-1}$ with aging at 185 °C for 13 h in an underaged Al–Cu–Mg–Ag alloy. This considerable expansion makes it possible to shrink or moreover close the fatigue cavity/crack by dynamic θ' phase precipitation on the concentrated area of deformation in the vicinity of fatigue cavity/crack.

7.3.3.3 Replenishment of Strengthening Phase by Dynamic Precipitation on Dislocation

Solute atoms such as Cu in Al alloys segregate to dislocations and facilitate precipitation there at ambient temperature during fatigue. Electron microscope observations showed that the dynamic precipitation of θ' phase on dislocations occurs in locations where the mobility of dislocations appears to have been impeded in the underaged Cu-containing Al alloy [8]. Such dynamic precipitation indicates that the dislocations are saturated with free solute atoms. The segregation of solute

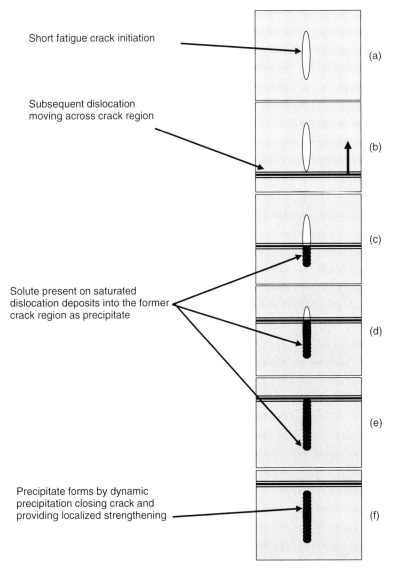

Short fatigue crack initiation

Subsequent dislocation moving across crack region

Solute present on saturated dislocation deposits into the former crack region as precipitate

Precipitate forms by dynamic precipitation closing crack and providing localized strengthening

(a)

(b)

(c)

(d)

(e)

(f)

Fig. 7.25 Proposed fatigue crack closure mechanism by the process of dynamic precipitation. Solute saturated along a dislocation is precipitated, thereby effectively closing the crack and providing localized strengthening [8].

atoms on dislocations and the dynamic precipitation there occur in intensively deformed locations and are expected to delay the fatigue cavity/crack initiation through immobilizing dislocations and localized hardening due to the dynamic precipitation.

7.3.4
Effect of Self-healing on Fatigue Properties of Al Alloy

Effect of solute Cu atoms on fatigue properties and microstructures was studied and evaluated from the viewpoint of self-healing of fatigue cavity/crack [8]. Extruded bars of an Al alloy with the composition of Al–5.6Cu–0.45 Mg–0.45Ag–0.3Mn–0.18Zr were heat-treated and fatigue tested. The bars of the alloy were solution treated in an air furnace for 6 h at 525 °C, cold water quenched, and aged in oil. The conditions of aging were (i) PA peakaged for 10 h at 185 °C, (ii) UA underaged for 2 h at 185 °C, followed by quenching to room temperature. Fatigue tests were conducted immediately after aging or storing at −4 °C.

Fatigue tests were carried out with high cycle reversed stress. The results are shown in Figure 7.26 [8], which demonstrates longer fatigue lives for the UA condition. The superiority of the UA condition becomes more marked with increasing fatigue life.

Fractography with SEM indicated that fatigue cracks nucleate at internal sites. Microstructural analysis with transmission electron microscopy (TEM) proved the dynamic precipitation of θ' phase associated with dislocations, suggesting that during fatigue the dislocations in the UA condition became saturated with free solute atoms.

Lumley *et al.* [8, 9] proposed the following mechanisms for the superiority of the UA condition in the Al alloy:

1. general reduction in dislocation mobility by segregation of Cu atoms and dynamic precipitation of θ' phase on dislocations.
2. Closing fine fatigue cavity/crack with delivery of solute Cu atoms to the cavity/crack sites, leading to delay in their initiation/propagation.

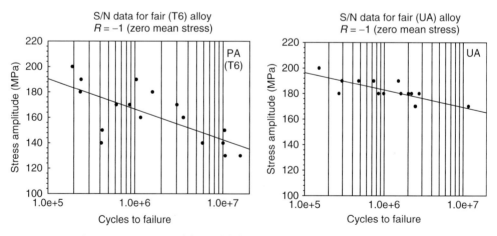

Fig. 7.26 Comparison of fatigue life between underaged (UA) and peakaged (PA) Al alloys. The UA material displays improved fatigue life compared to the PA treated alloy [8].

They also proposed a third possibility that fatigue cavity/crack initiation might be delayed by localized matrix hardening due to dynamic precipitation.

The development of self-healing Al alloys for actual use and the applications to airplanes are greatly expected. Latest report by Wanhil [45], however, suggested that the application of the Al alloy with the self-healing ability to airplanes is limited. The self-healing reaction in the concept is limited to internal cavity/crack, since its initiation at an external surface, such as in air environment, will be contaminated by adsorption of water vapor and oxygen onto the damage surfaces. This would prevent the self-healing reactions and certainly promote crack growth compared to fatigue in vacuum. More studies and improvements may be necessary for the actual applications of the self-healing of fatigue damage.

It is true that the proposal of the concept of self-healing for fatigue damage is very clear and original, and the experimental demonstration of the desired effect on fatigue properties has greatly stimulated the self-healing researches.

7.4
Summary and Remarks

A well-known self-healing in metallic materials is the protective passivation film of chrome oxide in stainless steels. The passivation film on surface is self-healed under oxidizing conditions, even if the chrome oxide is disrupted or destroyed, and continues to protect the surface from corrosion and oxidation. This type of surface self-healing has been providing stainless steels, Al alloys, and Ti alloys superior corrosion and oxidation resistance. In this self-healing process, the healing agent is O_2, which is delivered freely from the atmosphere, and the healing reaction is oxidation.

Compared with the surface self-healing for corrosion, self-healing of mechanical damage inside metallic materials is very difficult. Main reason is that there is no means for delivery of healing agents in metallic materials. The self-healing methods for polymers and composite materials have been well studied, and excellent methods have been proposed. These methods, however, are difficult to be applied to metallic materials. It is unrealistic to embed microcapsules or vascular networks filled with healing agents in metallic materials, and the adhesive as a self-healing agent is too weak to bond the cavities and cracks. For the development of self-healing creep cavity and fatigue cavity/crack in metallic materials, new ideas for delivery of healing agents to the damage sites and healing reactions for the fatal damages should be proposed.

Recently Shinya group [1–7] and Lumley group [8, 9] have proposed practical self-healing methods for creep cavity in heat resisting steels and fatigue cavity/crack in Al alloys, respectively. Their self-healing methods resemble each other and make use of solute atoms as healing agents. The solute atoms in heat resisting steels are delivered to surfaces of creep cavity through volume diffusion in the bulk, whereas the solute atoms in Al alloys are delivered to fatigue cavity/crack surfaces through pipe diffusion along dislocations. The delivered solute atoms segregate or precipitate on the cavity and crack surfaces. Main points of both self-healing methods are summarized and compared in Table 7.3.

Table 7.3 Main points of self-healing methods for creep cavity and fatigue cavity/crack.

	Shinya's group	Lumley's group
Materials damage	Creep cavities in heat resisting steels	Fatigue cavities/cracks in Al alloys
Healing agents	Solute B and solute N atoms	Solute Cu atoms
Delivery of healing agent	Volume diffusion of solute B and N atoms through bulk	Pipe diffusion of solute Cu atoms through dislocations
Healing reaction	Segregation of B atoms on creep cavity surfaces Precipitation of BN compound on creep cavity surfaces	Precipitation of Cu or Cu-containing phase on fatigue cavity/crack surfaces Segregation of Cu atoms and precipitation of Cu-containing phase on dislocations
Effect of reaction on healing	Suppression of creep cavity growth	Closure of fatigue cavities/cracks Suppression of localized deformation by pinning dislocation

The steels and the Al alloys with the self-healing of the damages could be produced easily only by additions of traces of B/Ce or small amount of Cu. Heat treatments for the materials are conventional one. Owing to the simple methods and low cost, the self-healing methods would be applied to a lot of materials and contribute to prevention against failures of plants and structures in the near future.

Acknowledgements

The author would like to thank Prof. Lumley for his permission to cite his figures from his papers, and also Mr Kyono and Dr Laha for their experimental practices and preparing figures.

References

1 Shinya, N., Kyono, J. and Laha, K. (2006) *Journal of Intelligent Material Systems and Structures*, **17**, 1127–33.

2 Laha, K., Kyono, J. and Shinya, N. (2007) *Philosophical Magazine*, **87**, 2483–2505.

3 Laha, K., Kyono, J. and Shinya, N. (2007) *Scripta Materialia*, **56**, 915–18.

4 Shinya, N. and Kyono, J. (2006) *Materials Transactions*, **47**, 2302–7.

5 Laha, K., Kyono, J., Sasaki, T., Kishimoto, S. and Shinya, N.

(2005) *Metallurgical and Materials Transactions A*, **36A**, 399–409.

6 Laha, K., Kyono, J., Kishimoto, S. and Shinya, N. (2005) *Scripta Metallurgica et Materialia*, **52**, 675–78.

7 Kyono, J. and Shinya, N. (2003) *Journal of The Society of Materials Science Japan*, **52**, 1211–16.

8 Lumley, R.N., O Donnell, K.G., Polmer, I.J. and Griffiths, J.R. (2005) *Materials Forum*, **29**, 256–61.

9 Lumley, R.N. and Polmer, I.J. (2007) "Proceedings of the 1st International Conference on Self Healing Materials", First International Conference on Self Healing Materials, Noordwijk aan Zee.

10 Viswanathan, R. (1985) *Journal of Pressure Vessel Technology*, **107**, 218–25.

11 Ashby, M.F., Gandhi, C.G. and Taplin, D.M.R. (1979) *Acta Metallurgica*, **27**, 699–729.

12 Shinya, N., Kyono, J. and Kushima, H. (2006) *ISIJ International*, **46**, 1516–22.

13 Shinya, N., Kyono, J., Tanaka, H., Murata, M. and Yokoi, S. (1983) *Tetsu-to-Hagané*, **69**, 1668–75.

14 Murata, M., Tanaka, H., Shinya, N. and Horiuchi, R. (1990) *Journal of The Society of Materials Science Japan*, **39**, 489–95.

15 Kyono, J., Shinya, N. and Horiuchi, R. (1992) *Tetsu-to-Hagané*, **79**, 604–10.

16 Shinya, N. and Keown, S.R. (1979) *Metal Science*, **13**, 89–93.

17 Speight, M.V. and Beeré, W. (1975) *Metal Science*, **9**, 190–91.

18 Evance, H.E. (1984) *Mechanisms of Creep Fracture*, Elsevier Applied Science Publishers, The Netherlands.

19 Needham, N.G. and Gladman, T. (1980) *Metal Science*, **14**, 64–72.

20 Nix, W.D., YU, K.S. and Wang, J.S. (1983) *Metallurgical Transactions A*, **14A**, 563–70.

21 White, C.L., Padget, R.A. and Swindeman, R.W. (1981) *Scripta Materialia*, **15**, 777–82.

22 Rhead, G.E. (1975) *Surface Science*, **247**, 207–21.

23 Delamare, F. and Rhead, G.E. (1971) *Surface Science*, **28**, 267–84.

24 Cosandey, F., Li, D., Sczerzenie, F. and Tien, J.K. (1983) *Metallurgical Transactions A*, **14A**, 611–21.

25 Hua, M., Garcia, C.J. and DeArdo, A.J. (1993) *Scripta Materialia*, **28**, 973–78.

26 Riedel, H. (1987) *Fracture at High Temperatures*, Springer, Berlin.

27 Good, S.H. and Nix, W.D. (1978) *Acta Metallurgica*, **26**, 739–52.

28 Holt, R.T. and Wallace, W. (1976) *International Material Reviews*, **21**, 1–24.

29 Minami, Y., Kimura, H. and Ihara, Y. (1986) *Materials Science and Technology*, **2**, 795–806.

30 Sourmail, T. (2001) *Materials Science and Technology*, **17**, 1–14.

31 Stulen, R.H. and Bastasz, R. (1979) *Journal of Vacuum Science and Technology*, **16**, 940–45.

32 Yoshihara, K., Tosa, M. and Nii, K. (1985) *Journal of Vacuum Science and Technology A*, **3**, 1804–808.

33 Nii, K. and Yoshihara, K. (1987) *Journal of Materials Engineering*, **9**, 41–50.

34 Mehrer, H. (eds) (1991) Landolt-Börnstein numerical data and functional relationships, in *Science and Technology, 3, 26, Diffusion in Metals and Alloys*, Springer-Verlag, pp. 195–96.

35 Klensnil, M. and Luká š, P. (1980) *Fatigue of Metallic Materials*, Elsevier Science Publishing, p. 59.

36 Polmer, I.J. and Bainbridge, I.F. (1959) *Philosophical Magazine*, **4**, 1293–1304.

37 Jannot, E., Mohles, V., Gottstein, G. and Thijsec, B. (2006) *Defect and Diffusion Forum*, **249**, 47–54.

38 Hautakangas, S., Schut, H., Zwaag, S., Castillo, P.E.J. and Dijk, N.H. (2007) Proceedings of the 1st International Conference on Self Healing Materials, First International Conference on Self Healing Materials, Noordwijk aan Zee.

39 Nagai, Y., Tang, Z., Hasegawa, M., Kanai, T. and Saneyasu, M. **(2001)** *Physical Review,* **63,** 134110.

40 Nagai, Y., Murayama, M., Tang, Z., Nonaka, T., Hono, K. and Hasegawa, M. **(2001)** *Acta Materialia,* **49,** 913–20.

41 Nagai, Y. and Hasegawa, M. **(2005)** *Materia Japan,* **44,** 667–73.

42 Nagai, Y. and Hasegawa, M. **(2001)** *Kinzoku,* **71,** 742–48.

43 Lumley, R.N. and Schaffer, G.B. **(2006)** *Scripta Materialia,* **55,** 207–10.

44 Hunsicker, H.Y. **(1980)** *Metallurgical Transactions A,* **11A,** 759–73.

45 Wanhil, R.J.H. **(2007)** Proceedings of the 1st International Conference on Self Healing Materials, First International Conference on Self Healing Materials, Noordwijk aan Zee.

8

Principles of Self-healing in Metals and Alloys: An Introduction

Michele V. Manuel

8.1
Introduction

Metals have played a critical role in the history of humankind. Its influence on human development is evident when studying the early tools and artifacts used for protection, shelter, and the sustainability of civilizations. In fact, key periods in the history of humankind have been named after the particular metals that dominated that era—Bronze and Iron ages. Metals continue to play a vital role in today's society. From technologically advanced components such as medical stents made from nickel and titanium, which actuate with a person's body temperature, to ultrahigh strength steels used in high performance automotive and aerospace applications, the need for high strength, light-weight and high temperature resistant materials has brought about another metal renaissance to humankind.

The pervasive use of materials such as metals is integrated so deeply into society that when they fail, the consequences can be devastating. To avoid endangering lives and repairing costly damages, an important design requirement that is often used is that the material does not fail during service. Failures involving fatigue, creep, and fracture can readily occur if materials are designed and used inappropriately. Additionally, even if a material is perfectly designed, it can still contain defects that are produced during processing or service, which may cause the material to degrade and possibly fail. Thus, developing a metal that can accommodate the stochastic nature of the service environment and materials properties is a daunting task. In applications where a designer needs to maximize reliability in structural materials, an ideal solution would be a self-healing metal. Nevertheless, in spite of the potential benefits and commercial importance of a self-healing structural metal, there has been very little research published in this area.

A common theme in the research that has been published in the area of self-healing metals is microstructure manipulation and control. Self-healing metals can be broadly classified as per their microstructural characteristics into

Self-healing Materials: Fundamentals, Design Strategies, and Applications. Edited by Swapan Kumar Ghosh
Copyright © 2009 WILEY-VCH Verlag GmbH & Co. KGaA, Weinheim
ISBN: 978-3-527-31829-2

two categories: liquid-assisted and solid-state healing. These categories describe the mechanism by which the self-healing metal transports matter to the damage site. As for the first category, the design of self-healing metals relying on liquid-assisted healing has centered on metal–matrix composites reinforced with shape memory alloy (SMA) wires in which the matrix partially liquefies at an elevated temperature. The microstructure is designed to maintain structural stability during healing by only allowing the liquid available at the crack surface to participate in damage remediation. The metals in the second category rely on the solid-state healing mechanism to exploit the strong driving force for solute diffusion to high-energy surfaces such as cracks or voids. Work in this area has primarily focused on aluminum alloys and steels. In both categories of self-healing metals, numerical and analytical modeling has been used to further understand the behavior and explore the mechanisms that give these metals their unique characteristics. This chapter highlights the approaches of several researchers who have developed metals with the ability to self-heal and/or self-repair.

8.2
Liquid-based Healing Mechanism

The most common design strategy used in self-healing polymeric materials is to embed a liquid phase healing agent within a solid matrix [1, 2]. This strategy can also be adopted in metals. Using a thermodynamic approach, the microstructure of the metal can be designed with a certain fraction of a low melting phase. Gradual heating of the metal causes partial liquefaction of the matrix, and thus the availability of a liquid healing agent for healing. Since the possibility of a liquid healing agent is always available, damage can be repaired multiple times over the entire life of the material. This mechanism allows for virtually limitless healing of fractures.

Although there are many benefits to a liquid-based healing mechanism (such as rapid healing kinetics), there is one design constraint that has limited the performance of these materials. This constraint relates to a limitation on the maximum allowable flaw size. Healing large flaws and defects is difficult since complete distribution of the healing agent relies on the forces. Large-scale damage can lead to the weakening of capillary forces and thus incomplete healing. One alternative is to increase the amount of healing agent, however, this can degrade the mechanical properties of the material. Another approach is to decrease the size of large-scale cracks by utilizing a crack closure mechanism such as SMA reinforcements.

A liquid-assisted SMA reinforced self-healing alloy composite has been developed using a thermodynamics-based systems design approach [3–6]. The schematic in Figure 8.1 illustrates the complete healing process. During operation, when a force is applied to the composite, which is large enough to initiate a matrix crack, crack nucleation and propagation occur. Owing to a limited interfacial strength at the

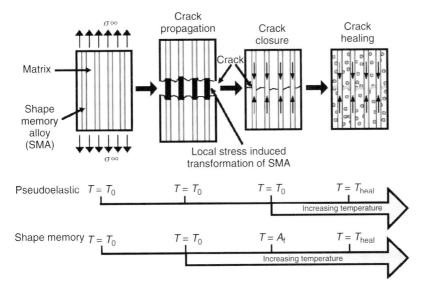

Fig. 8.1 A schematic of a metal–matrix composite reinforced by SMA wires undergoing a liquid-assisted self-healing process.

reinforcement interface and the high SMA wire strength relative to the matrix, the crack will cause interfacial debonding and crack bridging. These mechanisms serve to shield the crack, thereby increasing the overall toughness of the composite. Stretching of the bridged SMA wires occurs by stress-induced transformation, whereby the martensite variants locally reorient to accommodate the applied stress and minimize fiber pull-out.

When the composite is heated above the reversion temperature of the embedded SMA wires, a clamping force is applied by the SMA wires to provide crack closure. However, if the composite is reinforced with pseudoelastic SMA wires, crack closure would occur immediately following the release of the applied load. When the composite reaches its final healing temperature, the matrix will partially liquefy and cause crack filling. At this point, the SMA wires play another critical role by removing the plasticity produced during initial fracture as seen in Figure 8.2. The partial liquefaction of the matrix causes softening, which allows the SMA wire to exert enough force to reverse the plasticity and allow the composite to regain its original shape. Following healing and shape recovery, subsequent cooling allows the composite to regain its original tensile strength. The healing temperature is designed to be low to minimize the energy requirements for healing. In addition, controllable partial melting is designed to minimize liquid continuity in order to maintain the structural integrity of the matrix during healing.

Figure 8.3 illustrates the systems design approach utilized in the design of self-healing alloy composite. The hierarchical structure of the chart reflects the hierarchical nature of the composite. The property objectives are prioritized in the order of importance with memory healing being the most important followed by

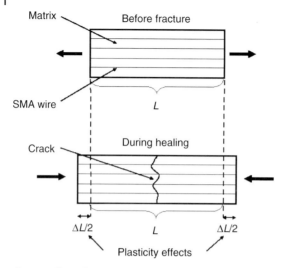

Fig. 8.2 Effect of plasticity on crack closure. The symbol ΔL represents the increase in length caused by plastic deformation of the matrix during fracture [5, 6].

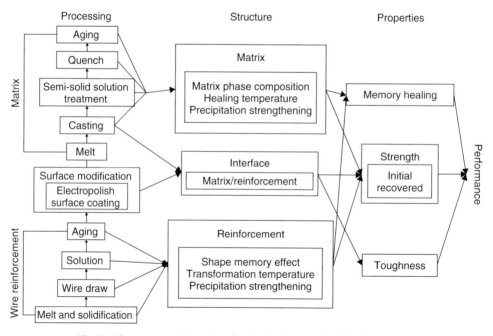

Fig. 8.3 The systems design chart for the liquid-assisted self-healing alloy composite [5, 6].

high strength, both initial and recovered, and high toughness. The hierarchy is also reflected in the vertical orientation of the chart, which outlines the processing steps from first (melting of the SMA wire) to last (matrix aging).

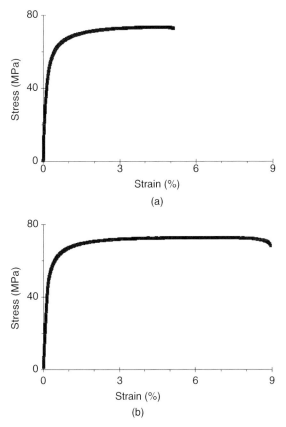

Fig. 8.4 Stress versus strain curve for the heat-treated Sn–21Bi (wt%) proof-of-concept (a) matrix and (b) composite with 1% volume fraction of SMA wires [5, 6].

Furthermore, the chart illustrates the major microstructural subsystems of interest that control the property objectives. The linkages represent key structure–property relationships that can be modeled and optimized for composite design. Meanwhile, the process–structure relationships can be predicted using science-based models and computational thermodynamics [5, 6].

Figure 8.4 displays a comparison of tensile engineering stress–strain curves of a proof-of-concept Sn–21Bi (wt%) matrix alloy and a Sn–21Bi (wt%) matrix alloy reinforced with 1% equiatomic NiTi SMA wires. The alloy reinforced with SMA wires displays a 73% increase in uniform ductility over the unreinforced matrix [5]. This increase in toughness has been attributed to crack deflection and interfacial debonding at the matrix/reinforcement interface. After complete matrix fracture, the composite was healed at 169 °C for 24 h, and then tensile tested again at room temperature. The specimens demonstrated a 95% tensile strength recovery. Optical micrographs of a cracked and then healed specimen are shown in Figure 8.5.

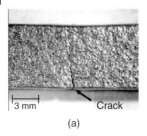

(a)

(b)

Fig. 8.5 Optical micrographs of a Sn–21Bi (wt%) proof-of-concept composite (a) after fracture and then (b) after healing. It is evident that the composite was able to heal a macroscopic crack [5, 6].

8.2.1
Modeling of a Liquid-assisted Self-healing Metal

A numerical model of the liquid-assisted self-healing alloy composite was developed by the Brinson group at Northwestern University [7] to study the thermomechanical interaction between the SMA wires and the matrix. The matrix behavior was modeled as a simple two-dimensional elastic–plastic material and the SMA wires were modeled using a one-dimensional SMA constitutive model [8]. The model can reproduce the constitutive behavior of the SMA wires at all temperature ranges. Using a user subroutine, the one-dimensional SMA model was integrated into the commercial ABAQUS finite element analysis code. ABAQUS is a software program that allows users to integrate their own user elements with on board software to solve problems by discretizing the model into a series of finite elements. The model allows users to input material parameters for each constituent composite phase to predict the behavior of the composite before fabrication.

The results from the simulation indicated a dramatic increase in the stress in the SMA wire during healing. Thus, to design a composite that prevents the maximum stress in the wire from exceeding the wire yield strength and causing damage, an analytical model was developed by Manuel and Olson [5, 6]. The model analyzes the relationships between the maximum stress in the wire during healing, the matrix strength, the strength of the SMA wires, and reinforcement volume fraction. It has been shown that the results obtainable through the simplified analytical model correlate well with the Brinson finite element model.

8.3
Healing in the Solid State: Precipitation-assisted Self-healing Metals

Diffusion occurs to minimize the energy of the system, bringing the system to equilibrium. The diffusion of atoms through a solid is typically slower than the diffusion of atoms through a liquid. However, solid-state diffusion can occur rather quickly along high-energy paths such as free surfaces, grain boundaries, and dislocations. In metals, the total grain boundary and dislocation area can be greater than the total surface area. It is this large area of high diffusivity paths in metals that makes this mechanism for mass transport an attractive option to deliver materials to a damage site for self-healing in the solid state.

8.3.1
Basic Phenomena: Age (Precipitation) Hardening

Many engineering alloys are age (precipitation) hardened to enhance their mechanical properties. Traditionally, these alloys are heat-treated at an elevated temperature and then rapidly cooled (quenched) to room temperature. This produces an alloy with a nonequilibrium microstructure that is supersaturated with solute atoms. During subsequent thermal treatment (aging), the supersaturated solid-solution equilibrates and decomposes to nucleate precipitates. The nucleation of precipitates typically occurs heterogeneously at nonequilibrium, high-energy defect sites, such as dislocations, grain boundaries, and free surfaces. In addition to providing ideal nucleation sites, the enhanced mobility of solute atoms along defect sites allows precipitates to grow rapidly as solute atoms are channeled to the particle [9]. Examples of high diffusivity paths are illustrated in Figure 8.6. This mechanism is traditionally used to form precipitates, which are designed to act as barriers to

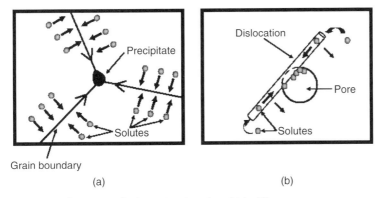

(a)   (b)

Fig. 8.6 An illustration of solute migration along high diffusivity paths and their precipitation on high-energy surfaces. The images show (a) solute diffusion along a grain boundary leading to a precipitate located at the triple point of three grains and (b) a mobile dislocation moving solutes to a pore.

dislocation motion, significantly increasing the strength of the alloy. However, it is this *in-situ* strengthening method that has recently been used in self-healing metals to produce precipitates at the surfaces of internal microcracks.

8.3.2
Self-healing in Aluminum Alloys

Recent work by Lumley and Polmear [10] has revealed a self-healing mechanism in underaged aluminum alloys by dynamic solute precipitation during creep and fatigue loading. In underaged alloys, the residual solute that has yet to undergo precipitation is available to precipitate at open volume defects under applied loading. According to Lumley and Schaffer [11], this mechanism is analogous to the precipitation-induced densification mechanism seen in Al–8Zn–2.5Mg–1Cu (in wt%) powder alloys. In these alloys, slow cooling from 620 °C produces a heterogeneous precipitation of the equilibrium η phase ($MgZn_2$) on open pore surfaces, thereby leading to an 8% reduction in porosity. Figure 8.7 displays a backscattered scanning electron microscope (SEM) micrograph of the Al–Zn–Mg–Cu alloy showing the precipitation of the η phase (white phase) on the surface of a pore.

Creep studies were performed on a series of aluminum alloys to investigate the performance between an alloy in the underaged and fully hardened (T6) condition. Results from these investigations revealed that the underaged alloy demonstrates a reduction in secondary creep rate as shown in Figure 8.8a and an increase in strain to failure (Figure 8.8b). This increase in creep resistance has been attributed to the presence of residual solutes forming solute atmospheres around dislocations (acting as a barrier to dislocation motion) and dynamic precipitation of the θ' phase on defect structures [10].

Fatigue studies at an *R*-value of −1 were conducted on underaged and fully hardened (T6) extruded Al–4Cu–0.3Mg–0.4Ag (wt%) alloys. After the aging treatment, the underaged and fully hardened alloys display a yield strength of 417 and 470 MPa, respectively [10]. Previous studies have shown that in precipitation strengthened aluminum alloys, increases in tensile strength correlates to decreases

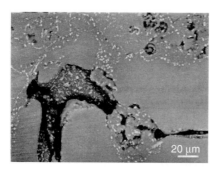

Fig. 8.7 Backscattered SEM image of η phase (white phase) precipitating on a pore in an Al–8Zn–2.5Mg–1Cu (wt%) alloy [11] (Reprinted with permission from Elsevier).

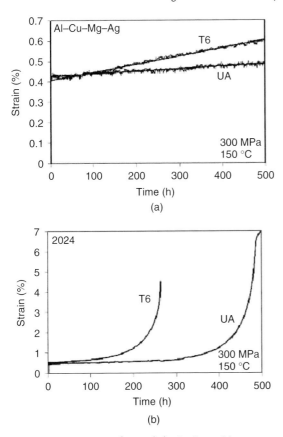

Fig. 8.8 A comparison of creep behavior in an (a)
Al−4Cu−0.3Mg−0.4Ag (wt%) and (b) 2024 aluminum alloy
in the underaged and fully hardened (T6) aluminum alloys
[10, 11] (Reprinted with permission from Elsevier).

in fatigue performance. Thus, fully hardened aluminum alloys demonstrate lower
fatigue strength than underaged alloys. It has been suggested that strain localiza-
tion causes precipitates situated on certain slip bands to be cut, as dislocations
move within these bands. The cutting causes the precipitates to become smaller
than the critical size needed for thermodynamic stability, thereby causing them
to dissolve. It is the dissolution of stable precipitates that leads to softening and a
decrease in fatigue properties [13].

Figure 8.9 shows the maximum stress versus number of cycles (fatigue S−N
curves) to failure of two aluminum alloys. The underaged alloy demonstrates a
greater number of cycles to failure over a range of stress levels. This increase in
fatigue strength has been attributed to the continual replenishment of θ' phase
precipitates on defect structures. As precipitates are being removed by dislocation
cutting, solutes dynamically precipitate causing dislocation pinning and healing of
the microstructure [10].

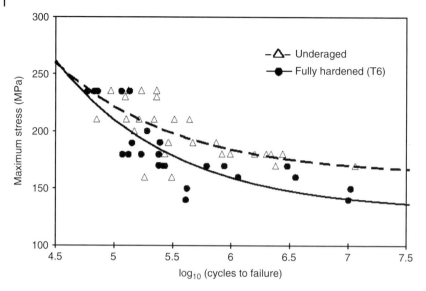

Fig. 8.9 A comparison of fatigue behavior from an Al–4Cu–0.3Mg–0.4Ag (wt%) alloy in the underaged and fully hardened (T6) condition. The curves represent an exponential fit to the data. Adapted from [10].

The precipitation-induced densification mechanism revealed by Lumley and Polmear [10] was further investigated in underaged commercial AA2024 aluminum alloys by Hautakangas *et al.* [14, 15] using positron lifetime spectroscopy. This technique uses positrons to evaluate strain-induced defects as a function of aging time. A positron is an electron antiparticle that can be trapped by open volume defects, negative ions, or solute aggregates with a higher positron affinity than the matrix. The work by Hautakangas *et al.* verifies Lumley's observation that dynamic precipitation is occurring on defect structures. They concluded that dynamic precipitation is assisted by strain-induced vacancies and pipe diffusion of solute atoms through the dislocation core (similar to the effect seen in Figure 8.6b). This allows the retained solute atoms to precipitate at internal open volume defects leading to self-healing.

In light of the work by Lumley and Polmear [10] on self-healing aluminum alloys, a recent study has highlighted the need to characterize the self-healing behavior of metals under conditions similar that seen in service. Studies by Wanhill [16] have shown that in aerospace aluminum alloys, fatigue cracks are frequently initiated at external surfaces. These surfaces are continually exposed to corrosive environments reducing the ability for an externally imitated fatigue crack to self-heal. It is also suggested that although there may be experimental evidence of solute migration to free surfaces causing filling of internal cracks and voids, this phenomenon cannot solely explain the change in fatigue behavior. It is a suggested notion that dislocation drag by solute atmosphere and dislocation pinning by dynamic precipitation are mechanisms that are independent of self-healing and may contribute to the increased fatigue strength of aluminum alloys.

8.3.3
Self-healing in Steels

A metal that has demonstrated the potential for self-healing creep cavities is steel. High services temperatures can significantly reduce the creep strength and ductility of steel. Creep fracture occurs when cavities nucleate on the grain boundaries, grow, and then coalesce. Studies by Shinya *et al.* and Laha *et al.* [17, 18] have shown that creep cavitation is accelerated by the presence of sulfur. It has also been shown that the removal of sulfur promotes the diffusion of boron to the creep cavity to precipitate boron nitride on the creep cavity surface. Studies on type 304 austenitic stainless steels have demonstrated continuous boron nitride precipitation causing self-healing of the microstructure. It has been suggested that the precipitation of boron nitride plays a dual role in the enhancement of creep performance. The first role is the filling of creep cavities and the second role is the suppression of creep cavity growth through the alteration of its surface diffusion characteristics [19]. The suppression of creep cavity formation can be seen in Figure 8.10. A detailed discussion on self-healing in steel can be found in Chapter 7.

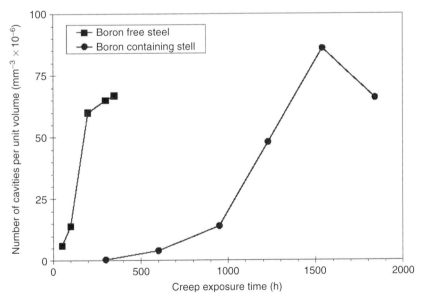

Fig. 8.10 Change in the number of cavities per unit volume in a boron-free and boron-containing 347 type austenitic stainless steels exposed to 78 MPa at 750 °C. Adapted from [18].

8.3.4
Modeling of Solid-state Healing

Modeling can provide a deeper insight into the dynamic behavior of healing in the solid state. By simulating conditions that are not so easy to perform experimentally, modeling can provide a glimpse into the dramatic microstructural changes that can occur during healing.

Early studies of solid-state crack healing in metals utilized molecular dynamics simulations to investigate the healing of internal microcracks in pure aluminum and copper [20–22]. These investigations focused on the effect of dislocation emission from a crack tip and dislocation motion on crack healing. It was determined that the application of a compressive stress or heating during healing could increase healing efficiency. Additionally, it was also noted that the presence of a dislocation around a microcrack would further promote healing. Later, this methodology was used to study crack healing in body-centered cubic iron [23], which suggested the operation of mechanisms similar to that seen earlier in aluminum and copper.

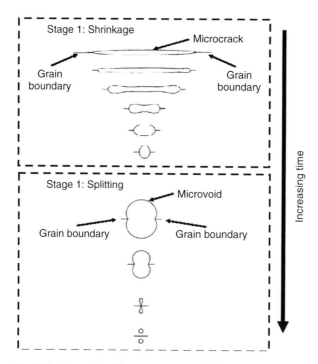

Fig. 8.11 Schematic of the two-stage morphological evolution of a microcrack lying along a grain boundary being subjected to a far-field external stress. The initial aspect ratio (length/height) of the microcrack equals 20. Adapted from [24].

Numerical simulations using a mechanics and thermodynamics approach have been used to predict the morphological evolution of micron- and submicron-sized internal defects during healing. Using finite element analysis, Huang *et al.* [24] revealed that there are two stages involved in the healing of internal ellipsoid shaped microcracks healed under externally applied pressure. The first stage involves shrinking of the microcrack owing to grain boundary diffusion of atoms to the crack surface, which eventually leads to the recession and blunting of the crack tips. The next stage is driven by surface diffusion and involves splitting of the microcrack into cylindrical void channels. Figure 8.11 illustrates a 2D simulation of the morphological evolution of a microcrack lying along a grain boundary under external far-field stress. However, since these void channels are highly unstable, Wang and Li [25, 26] have taken the next step to analyze the void shrinkage rate using a thermodynamic approach, taking into account the void surface, grain boundary, and elastic stored energy in the metal. Their analysis establishes a maximum external applied pressure, which if exceeded causes void collapse prior to complete healing and the retardation of crack healing due to increasing internal gas pressure (which can be avoided if healing is performed in vacuum [27]). These theoretical analyses of microcrack healing correlate well with experimental observations [27, 28].

8.4
Conclusions

There has always been an interest in the development of metals that can demonstrate healing properties, thus creating a large driving force to advance the field of self-healing metals. Already metal–matrix composites utilizing a liquid-based healing mechanism have demonstrated the ability to recover tensile strength after large-scale damage while metals using the solid-state-assisted healing have demonstrated high levels of damage tolerance and removal of internal cracks and voids. Both categories of self-healing metals have their benefits and limitations. It seems that for damage healing at all length scales and under various environmental conditions, a combination of liquid-assisted and solid-state mechanisms should perhaps be used for complete self-healing. However, it is important to note that this research area is still in its infancy and although impressive, the healing ability of metals is far from the characteristic healing behavior seen in biological materials. Nonetheless, the potential benefits of a self-healing metal are numerous and this research area will continue to grow to bridge the gap.

Acknowledgments

The author would like to gratefully acknowledge Anil Sachdev and Gregory Olson for their help and advice.

References

1 Kessler, M.R. (**2007**) *Proceedings of the Institution of Mechanical Engineers Part G-Journal of Aerospace Engineering*, **221** (G4), 479–95.

2 Keller, M.W., Blaiszik, B.J., White, S.R. and Sottos, N.R. (**2007**) Recent advances in self-healing materials systems, in *Adaptive Structures: Engineering Applications* (eds D. Wagg, I. Bond, P. Weaver and M. Friswell), John Wiley & Sons, West Sussex, pp. 247–60.

3 Files, B.S. (**1997**) *Design of a Biomimetic Self-Healing Superalloy Composite*, Dissertation, Northwestern University.

4 Files, B. and Olson, G. (**1997**) "Terminator 3: biomimetic self-healing alloy composite", in *SMST-97: Proceedings of the Second International Conference on Shape Memory and Superelastic Technologies* (eds A. Pelton, D. Hodgson, S. Russell and T. Duerig), The International Organization on Shape Memory and Superelastic Technologies (SMST), Pacific Grove, pp. 281–86.

5 Manuel, M.V. (**2007**) *Design of a Biomimetic Self-Healing Alloy Composite*, Dissertation, Northwestern University.

6 Manuel M.V. and Olson, G.B. (**2007**) Biomimetic self-healing metals , in *Proceedings of the First International Conference on Self Healing Materials*, Series in Materials Science, Vol. 100, Springer, Noordwijk aan Zee, 18–20 April 2007.

7 Burton, D.S., Gao, X. and Brinson, L.C. (**2006**) *Mechanics of Materials*, **38**, 525–37.

8 Brinson, L.C. (**1993**) *Journal of Intelligent Material Systems and Structures*, **4**, 229–42.

9 Shewmon, P.G. (**1970**) Diffusion, in *Physical Metallurgy*, 2nd edn (ed R.W. Cahn), North-Holland Publishing Company, Amsterdam.

10 Lumley, R.N. and Polmear, I.J. (**2007**) "Advances in self-healing metals". *Proceedings of the First International Conference on Self Healing Materials*, Series in Materials Science, Vol. 24, Springer, Noordwijk aan Zee, 18–20 April 2007.

11 Lumley, R.N., Morton, A.J. and Polmear, I.J. (**2002**) *Acta Materialia*, **50**, 3597–3608.

12 Lumley, R.N. and Schaffer, G.B. (**2006**) *Scripta Materialia*, **55**, 207–10.

13 Polmear, I.J. (**1981**) *Light Alloys: Metallurgy of the Light Metals*, Edward Arnold, London.

14 Hautakangas, S., Schut, H., van der Zwaag, S., Rivera Diaz del Castillo, P.E.J. and van Dijk, N.H. (**2007**) *Physical Status Solidi C*, **4** (10), 3469–72.

15 Hautakangas, S., Schut, H., van der Zwaag, S. Rivera Diaz del Castillo, P.E.J. and van Dijk, N.H. (**2007**) The role of the aging temperature on the self-healing kinetics in an underaged aa2024 aluminum alloy . *Proceedings of the First International Conference on Self Healing Materials*, Series in Materials Science, Vol. 100, Springer, Noordwijk aan Zee, 18–20 April 2007.

16 Wanhill, R.J.H. (**2007**) "Fatigue crack initiation in aerospace aluminum alloys, components and structures", *First International Conference on Self-Healing Materials*, Springer, Noordwijk aan Zee.

17 Shinya, N., Kyono, J., Laha, K. and Masuda, C. (**2007**) Self healing of creep damage through autonomous boron segregation and boron nitride precipitation during high temperature use of austenitic stainless steels . *Proceedings of the First International Conference on Self Healing Materials*, Series in Materials Science, Vol. 100, Springer, Noordwijk aan Zee, 18–20 April 2007.

18 Laha, K., Kyono, J., Sasaki, T., Kishimoto, S. and Shinya, N. (**2005**) *Metallurgical and Materials Transactions A*, **36**, 399–409.

19 Shinya, N., Kyono, J. and Laha, K. (**2006**) *Journal of Intelligent Material Systems and Structures*, **17**, 1127–33.

20 Li, S., Gao, K.W., Qiao, L.J., Zhou, F.X. and Chu, W.Y. (**2001**) *Computational Materials Science*, **20**, 143–50.

21 Shen, L., Kewei, G., Lijie, Q. and Wuyang, C. (**2000**) *Acta Mechanica Sinica*, **16** (4), 366–73.

22 Zhou, G., Gao, K., Qiao, L., Wang, Y. and Chu, W. (**2000**) *Materials Science and Engineering*, **8**, 603–9.

23 Wei, D., IIan, J., Kict Tieu, A. and Jiang, Z. (**2004**) *Scripta Materialia*, **51**, 583–87.

24 Huang, P., Li, Z. and Sun, J. (**2002**) *Metallurgical and Materials Transactions A*, **33**, 1117–24.

25 Wang, H. and Li, Z. (**2003**) *Metallurgical and Materials Transactions A*, **34**, 1493–500.

26 Wang, H. and Li, Z. (**2004**) *Journal of Applied Physics*, **95** (11), 6025–31.

27 Zhang, H.L. and Sun, J. (**2004**) *Materials Science and Engineering A*, **382**, 171–80.

28 Wei, D., Han, J., Jiang, Z.Y., Lu, C. and Tieu, A.K. (**2006**) *Journal of Materials Processing Technology*, **117**, 233–37.

9

Modeling Self-healing of Fiber-reinforced Polymer–matrix Composites with Distributed Damage

Ever J. Barbero, Kevin J. Ford and Joan A. Mayugo

9.1
Introduction

Composite materials are formed by the combination of two or more distinct materials to form a new material with enhanced properties [1]. Recently, self-healing polymers and composites have been proposed. One system in particular incorporates the use of ruthenium catalyst and urea-formaldehyde microcapsules filled with dicyclopentadiene (DCPD) [2]. Barbero and Ford [3] accomplished self-healing of glass fiber-reinforced epoxy laminates by intralaminar dispersion of healing agent and catalyst. The healing agent, DCPD, is encapsulated and then dispersed in the epoxy resin during hand layup. The catalyst is also encapsulated and dispersed in a similar manner. Vacuum bagging technique is used to consolidate the samples that are cured at room temperature. Then, experiments are performed to reveal damage, plasticity, and healing of laminates under cyclic load.

Barbero *et al.* [4] developed a continuous damage and healing mechanics (CDHM) model to predict the effects of damage and subsequent self-healing as a function of load history. In Ref [4], damage and healing are represented in separate thermodynamic spaces. The damage portion of the model has been extensively identified and verified with data available in the literature [5, Chapter 8]. The self-healing portion of the previous model could not be identified or verified because of lack of experimental data for laminates undergoing distributed damage (e.g. microcracking). Until recently, data existed only for fracture toughness recovery due to healing of macrocracks [2, 6–8]. Then, Barbero and Ford [3] conducted a comprehensive experimental study to identify the healing portion of the model. Their experimental evidence suggests that simplification of the earlier model is possible, thus motivating the model presented in this chapter. Specifically, a damage/healing model where both effects are described in a single thermodynamic space is presented, thus simplifying considerably the earlier model.

Self-healing Materials: Fundamentals, Design Strategies, and Applications. Edited by Swapan Kumar Ghosh
Copyright © 2009 WILEY-VCH Verlag GmbH & Co. KGaA, Weinheim
ISBN: 978-3-527-31829-2

The continuum damage-healing mechanics model proposed herein consists of a continuum damage mechanics model extended to account for healing effects. The model must be identified with experimental data from unidirectional or cross-ply laminates. Therefore, a methodology for identification of the damage and healing parameters of the model is described. Once identified, the model is capable of predicting damage and healing evolution of other laminate stacking sequence (LSS). Prediction of damage, healing, damage hardening, and hardening recovery upon healing are accomplished.

The model presented herein is likely to work, with modifications, for other healing processes such as geological rock densification [9], self-healing of concrete [10, 11], self-healing of ceramic materials [12, 13], bone remodeling, wounded skin regeneration [14–16], and compaction of crushed rock salt [9].

9.2
Damage Model

In this chapter, damage is represented by its effects on residual stiffness and strength. No attempt is made to identify actual microstructural modes of damage. However, various components of the model, namely, the damage variable adopted and several simplifying assumptions are rooted in experimental evidence. The model is composed of three main ingredients, the damage variable, the free-energy potential, and the damage evolution equations.

9.2.1
Damage Variable

Damage represents distributed, irreversible phenomena that cause stiffness and strength reductions. The choice of damage variable has a direct impact on the number of material parameters required to describe the phenomena and the accuracy of the model predictions. In this work, damage is represented by a second-order tensor $\mathbf{D}$. For convenience, the integrity tensor is defined as $\Omega = \sqrt{\mathbf{I} - \mathbf{D}}$, where $\mathbf{I}$ is the second-order identity tensor.

Experimental evidence suggests that damage in the form of microcracks, delamination, fiber break, and so on, occur in planes parallel to the principal material directions. Therefore, it is assumed that the principal directions of damage n_1, n_2, n_3 coincide with the material orientations x_1, x_2, x_3. In this case, the damage tensor can be described by its three eigenvalues d_1, d_2, d_3. Such simplification allows us to think of the eigenvalues as a representation of fictitious, equivalent system of cracks that represent damage in the longitudinal, transverse, and thickness directions, respectively.

The model is developed taking two configurations into account: damaged and effective. In the damaged (actual) configuration, the material is subjected to nominal stress and undergoes damage $\mathbf{D}$, which results in reduced stiffness $\mathbf{C}(\mathbf{D})$. The effective configuration is a fictitious configuration where an increased effective stress $\bar{\sigma}$ acts upon a fictitious material having undamaged (virgin) elastic stiffness $\overline{\mathbf{C}}$. According to the energy equivalence hypothesis ([5, Chapter 8]), the elastic energy in the actual and effective configurations is identical.

The transformation of stress and strain between effective and damaged configurations is accomplished by

$$\bar{\sigma}_{ij} = M_{ijkl}^{-1}\sigma_{kl}; \quad \bar{\varepsilon}_{ij}^e = M_{ijkl}\varepsilon_{kl}^e \tag{9.1}$$

where an over-bar indicates that the quantity is evaluated in the effective configuration, the superscript e denotes quantities in the elastic domain, and **M** is the *damage effect tensor* defined as

$$M_{ijkl} = \frac{1}{2}(\Omega_{ik}\Omega_{jl} + \Omega_{il}\Omega_{jk}) \tag{9.2}$$

The stress–strain relationship in the effective configuration is simply that of a linearly elastic material with virgin properties, given by

$$\bar{\sigma}_{ij} = \overline{C}_{ijkl}\bar{\varepsilon}_{kl}^e; \quad \bar{\varepsilon}_{ij}^e = \overline{C}_{ijkl}^{-1}\bar{\sigma}_{kl} = \overline{S}_{ijkl}\bar{\sigma}_{kl} \tag{9.3}$$

The constitutive equation in the damaged configuration is obtained by substituting Equations 9.3 into Equations 9.1,

$$
\begin{aligned}
\sigma_{ij} &= M_{ijkl}\bar{\sigma}_{kl} = M_{ijkl}\overline{C}_{klrs}\bar{\varepsilon}_{rs}^e, & \varepsilon_{ij}^e &= M_{ijkl}^{-1}\bar{\varepsilon}_{kl}^e = M_{ijkl}^{-1}\overline{S}_{klrs}\bar{\sigma}_{rs}, \\
\sigma_{ij} &= M_{ijkl}\overline{C}_{klrs}M_{rstu}\varepsilon_{tu}^e, & \varepsilon_{ij}^e &= M_{ijkl}^{-1}\overline{S}_{klrs}M_{ijkl}^{-1}\sigma_{tu}, \\
\sigma_{ij} &= C_{ijkl}\varepsilon_{kl}^e, & \varepsilon_{ij}^e &= S_{ijkl}\sigma_{kl}
\end{aligned}
\tag{9.4}
$$

where by virtue of **M** being symmetric and using the energy equivalence hypothesis, the constitutive tensors **C** and **S** in the damaged domain are symmetric tensors given by

$$C_{ijkl} = M_{ijkl}\overline{C}_{klrs}M_{rstu}; \quad S_{ijkl} = M_{ijkl}^{-1}\overline{S}_{klrs}M_{rstu}^{-1} \tag{9.5}$$

9.2.2
Free-energy Potential

The constitutive equations are derived from thermodynamic principles. The Helmholtz free energy includes the elastic energy and additional terms to represent the evolution of the internal parameters as follows:

$$\psi = \varphi(\boldsymbol{\varepsilon}, \boldsymbol{\varepsilon}^P, D) - c_1^d\left[c_2^d\exp\left(\frac{\delta}{c_2^d}\right)\right] - c_1^p\left[c_2^p\exp\left(\frac{\delta}{c_2^p}\right)\right] \tag{9.6}$$

where $\boldsymbol{\varepsilon}$, $\boldsymbol{\varepsilon}^P$ are the elastic and plastic strain tensors, respectively; p, δ are the hardening variables, and $c_1^d, c_2^d, c_1^p, c_2^p$, are material parameters used to adjust the model to experimental data.

The following thermodynamic state laws can be obtained by satisfying the Clausius–Duhem inequality [17], thus assuring nonnegative dissipation

$$\sigma_{ij} = \frac{\partial \psi}{\partial \varepsilon_{ij}^e} = C_{ijkl}(\varepsilon_{kl} - \varepsilon_{kl}^p) = C_{ijkl}\varepsilon_{kl}^e \tag{9.7}$$

$$Y_{ij} = -\frac{\partial \psi}{\partial D_{ij}} = -\frac{1}{2}(\varepsilon_{kl} - \varepsilon_{kl}^p)\frac{\partial C_{klpq}}{\partial D_{ij}}(\varepsilon_{pq} - \varepsilon_{pq}^p) = -\frac{1}{2}\varepsilon_{kl}^e\frac{\partial C_{klpq}}{\partial D_{ij}}\varepsilon_{pq}^e \tag{9.8}$$

$$\gamma(\delta) = -\frac{\partial \psi}{\partial \delta} = c_1^d\left[\exp\left(\frac{\delta}{c_2^d}\right) - 1\right] \tag{9.9}$$

$$R(p) = -\frac{\partial \psi}{\partial p} = c_1^p\left[\exp\left(\frac{p}{c_2^p}\right) - 1\right] \tag{9.10}$$

where σ, $\mathbf{Y}$, γ, and R are the thermodynamic forces associated with the internal state variables ε, $\mathbf{D}$, p, and δ.

The thermodynamic forces can be written explicitly in terms of stress [5, Appendix 2] as

$$Y_{11} = \frac{1}{\Omega_1^2}\left(\frac{\bar{S}_{11}}{\Omega_1^4}\sigma_1^2 + \frac{\bar{S}_{12}}{\Omega_1^2\Omega_2^2}\sigma_2\sigma_1 + \frac{\bar{S}_{13}}{\Omega_1^2\Omega_3^2}\sigma_3\sigma_1 + \frac{2\bar{S}_{55}}{\Omega_1^2\Omega_3^2}\sigma_5^2 + \frac{2\bar{S}_{66}}{\Omega_1^2\Omega_2^2}\sigma_6^2\right)$$

$$Y_{22} = \frac{1}{\Omega_2^2}\left(\frac{\bar{S}_{22}}{\Omega_2^4}\sigma_2^2 + \frac{\bar{S}_{12}}{\Omega_2^2\Omega_1^2}\sigma_2\sigma_1 + \frac{\bar{S}_{23}}{\Omega_2^2\Omega_3^2}\sigma_3\sigma_2 + \frac{2\bar{S}_{44}}{\Omega_2^2\Omega_3^2}\sigma_4^2 + \frac{2\bar{S}_{66}}{\Omega_2^2\Omega_1^2}\sigma_6^2\right)$$

$$Y_{33} = \frac{1}{\Omega_3^2}\left(\frac{\bar{S}_{33}}{\Omega_3^4}\sigma_3^2 + \frac{\bar{S}_{13}}{\Omega_3^2\Omega_1^2}\sigma_3\sigma_1 + \frac{\bar{S}_{23}}{\Omega_3^2\Omega_2^2}\sigma_3\sigma_2 + \frac{2\bar{S}_{44}}{\Omega_3^2\Omega_2^2}\sigma_4^2 + \frac{2\bar{S}_{55}}{\Omega_3^2\Omega_1^2}\sigma_5^2\right)$$

$$\tag{9.11}$$

9.2.3
Damage Evolution Equations

The evolution of the internal variables is defined as follows. First, damage initiation is controlled by a damage function, such as

$$g^d = \sqrt{\widehat{Y}^N : J : \widehat{Y}^N} + \sqrt{\widehat{Y}^S : B : \widehat{Y}^S} - [\gamma(\delta) + \gamma_0] \tag{9.12}$$

where γ_0 is the damage threshold and defines the hardening function. For plane stress, and taking into account the symmetry of the thermodynamic force tensor $\mathbf{Y}$, the damage surface is given by Equation 9.12 with parameters A_1, A_2, B_1, B_2, J_1, J_2.

The thermodynamic force tensor is assumed to be separable into a component $\mathbf{Y}^N$ arising from normal strains and a component $\mathbf{Y}^S = \mathbf{Y} - \mathbf{Y}^N$ arising from shear strains

$$\widehat{Y}^N = \widehat{A} : Y^N$$

$$\widehat{A}_{ijkl} = I_{ijkl} + (A_m - 1)\delta_{im}\delta_{jm}\delta_{km}\delta_{lm}\frac{\left(1 - \frac{\varepsilon_m}{|\varepsilon_m|}\right)}{2} \tag{9.13}$$

with $m = 1, 2, 3$. The values of ϵ_m are the values of the three normal strains in the principal material directions. The three values A_m represent the relation between damage thresholds in uniaxial compression and uniaxial extension in the principal material directions. The coefficients in the diagonal, fourth order, positive definite tensors $\mathbf{A}$, $\mathbf{B}$, $\mathbf{J}$ are calculated from available experimental data for unidirectional or cross-ply laminates as explained in Ref. [5, Chapter 9], while the parameters are determined by using experimental in-plane shear stress–strain data.

On the other hand, the plastic strain evolution is modeled by classical plasticity formulation [5, Chapter 9] and an associate flow rule. For the yield surface, a three-dimensional Tsai–Wu criterion shape is chosen due to its ability to represent different behavior among the different load paths in stress space. Plastic strains and damage effects are coupled by formulating the plasticity model in effective stress space. Therefore, the yield surface is a function of the thermodynamic forces $\bar{\sigma}$, R in the effective configuration as follows:

$$g^p = \sqrt{f_{ij}\bar{\sigma}_i\bar{\sigma}_j + f_i\bar{\sigma}_j} - [R(p) + R_0] \tag{9.14}$$

where $i = 1, 2, 6$, R_0 is the yield stress, and R is defined by the hardening law. The coefficients f_i, f_{ij} are obtained from the strength properties of unidirectional or cross-ply laminates in terms of the lamina strength values as follows:

$$f_1 = \frac{1}{\bar{F}_{1t}} - \frac{1}{\bar{F}_{1c}}; \qquad f_2 = \frac{1}{\bar{F}_{2t}} - \frac{1}{\bar{F}_{2c}}; \qquad f_3 = \frac{1}{\bar{F}_{3t}} - \frac{1}{\bar{F}_{3c}};$$

$$f_{11} = \frac{1}{\bar{F}_{1t}\bar{F}_{1c}}; \qquad f_{22} = \frac{1}{\bar{F}_{2t}\bar{F}_{2c}}; \qquad f_{33} = \frac{1}{\bar{F}_{3t}\bar{F}_{3c}};$$

$$f_{44} = \frac{1}{\bar{F}_4^2}; \qquad f_{55} = \frac{1}{\bar{F}_5^2}; \qquad f_{66} = \frac{1}{\bar{F}_6^2};$$

$$f_{23} \cong -\frac{1}{2}(f_{22}f_{33})^{1/2}; \qquad f_{13} \cong -\frac{1}{2}(f_{11}f_{33})^{1/2}; \qquad f_{12} \cong -\frac{1}{2}(f_{11}f_{22})^{1/2} \tag{9.15}$$

The parameters F_{it}, F_{ic}, and F_i are the effective strength values. That is, the strength values in effective configuration. They are defined as

$$\bar{F}_{1t} = \frac{F_{1t}}{\Omega_{1t}}; \qquad \bar{F}_{2t} = \frac{F_{2t}}{\Omega_{2t}}; \qquad \bar{F}_{3t} = \frac{F_{3t}}{\Omega_{3t}};$$

$$\bar{F}_{1c} = \frac{F_{1c}}{\Omega_{1c}}; \qquad \bar{F}_{2c} = \frac{F_{2c}}{\Omega_{2c}}; \qquad \bar{F}_{3c} = \frac{F_{3c}}{\Omega_{3c}};$$

$$\bar{F}_4 = F_4\frac{\bar{G}_{12}}{G_{12}^*}\frac{\bar{G}_{13}}{G_{13}^*}; \qquad \bar{F}_5 = F_5\frac{\bar{G}_{13}}{G_{13}^*}; \qquad \bar{F}_6 = F_6\frac{\bar{G}_{12}}{G_{12}^*} \tag{9.16}$$

where the parameters F_{it} and F_{ic} (with $i = 1, 2, 3$), and F_i (with $i = 4, 5, 6$) are the strength values in tension, compression, in-plane shear, and out-of-plane shear for a composite lamina. G_i^* and G_i (with $i = 12, 13$) are the damaged shear modulus and the undamaged shear modulus, respectively. These values are tabulated in the literature or they can be easily obtained following standardized test methods [3, 18].

The evolution of internal variables is defined by flow rules

$$\dot{\bar{\varepsilon}}_{ij}^{p} = \dot{\lambda}^{p}\frac{\partial g^{p}}{\partial \bar{\sigma}_{ij}}; \quad \dot{p} = \dot{\lambda}^{p}\frac{\partial g^{p}}{\partial R}$$

$$\dot{D}_{ij} = \dot{\lambda}^{d}\frac{\partial g^{d}}{\partial Y_{ij}}; \quad \dot{\delta} = \dot{\lambda}^{d}\frac{\partial g^{d}}{\partial \gamma} \tag{9.17}$$

in terms of plastic strain and damage multipliers $\dot{\lambda}^{p}$, $\dot{\lambda}^{d}$. These are found by using a return mapping algorithm [5, Chapter 9].

9.3
Healing Model

Healing represents the extent of repair of distributed damage. Similar to damage, healing is represented by second-order tensor **H**. Since healing can only heal existing damage, healing is represented by a diagonal tensor with principal directions aligned with those of the damage tensor **D**. The principal values h_1, h_2, h_3 represent the area recovery normal to the principal directions, which are aligned with the material directions x_1, x_2, x_3 in the material coordinate system. As a consequence of the particular definition of damage and healing used herein, the healing model presented uses the same thermodynamic space for damage and healing. That is, both damage and healing are defined in the space of thermodynamic damage forces **Y**.

In Refs [7, 19], the crack-healing efficiency is defined as the percentage recovery of fracture toughness measured by tapered double cantilever beam (TDCB). In this chapter, it is postulated that the healing tensor is proportional to the damage tensor

$$h_i = \eta_i d_i \tag{9.18}$$

The proportionality constant is the efficiency of the healing system. Furthermore, healing represents recovery of damage. Therefore, the principal values of the healed–damage tensor are given by

$$d_i^h = d_i - h_i = d_i(1 - \eta) \tag{9.19}$$

Experimentally, it is possible to measure the following (Figure 9.1):
- the virgin moduli $\overline{G}_i$ from the initial slope of the first cycle of loading;
- the damaged moduli G_i^{d} from the unloading portion of the first cycle; and
- the healed moduli G_i^h from the loading portion of the second cycle (after healing).

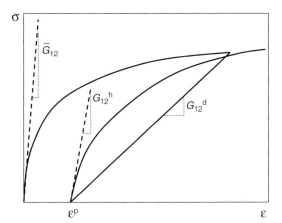

Fig. 9.1 Definition of measurable, relevant variables.

From these data, the efficiency is defined here as

$$\eta_i = \frac{G_i^h - G_i^d}{\overline{G}_i - G_i^d}$$ (9.20)

For composites reinforced by strong fibers, the self-healing system is incapable of healing fiber damage, which results in $\eta_1 = 0$. Once the self-healing polymer is released, it travels by capillary action and penetrates all the microcracks regardless of orientation, and thus $\eta_2 = \eta_3 = \eta$.

Once the material damages, higher levels of thermodynamic damage force **Y** are required to produce more damage. Higher thermodynamic damage force requires higher stress and correspondingly higher applied strain. This process, which is called *damage hardening*, is represented in the model by a state variable, the hardening parameter δ, which is a monotonically increasing function that provides a threshold below which no further damage can occur. A secondary effect of healing is to reduce the hardening threshold. That is, after healing, further damage can occur below the previous damage hardening threshold because some of the damage has been repaired. The model proposed herein captures this behavior by the following reduction in the damage hardening parameter

$$\delta^h = \delta(1 - \eta)$$ (9.21)

When the material is completely repaired, one has $\eta = 1$, $\delta = 0$, $\gamma = 0$ in the hardening law and the threshold for damage returns to the value for the virgin material. Thus, the material will begin to damage at the same threshold load as that of the virgin material. When the healing agent is exhausted, $\eta = 0$ and the material is just a damaging material with $\delta^h = \delta$; $\gamma \neq 0$ in the damage hardening law, while the threshold for damage continues to increase, thus representing the normal hardening behavior of a purely damaging material.

9.4
Damage and Plasticity Identification

For damage without healing, the damage parameters are A_1, A_2, B_1, B_2, J_1, J_2, γ_0, c_1^d, c_2^d. Plasticity entails additional parameters as follows: f_i, f_{ij}, R_0, c_1^p, c_2^p. All these parameters depend on the following material properties, which are experimentally determined using standard testing methods:

- stiffness values (E_1, E_2, G_{12})
- strength values $(F_{1t}, F_{1c}, F_{2c}, F_{2t}, F_6)$
- critical damage values (d_{1t}, d_{1c}, d_{2t})
- damaged shear moduli at imminent failure (G_{12}^*)
- in-plane shear plastic threshold (F_6^{EP})
- in-plane shear damage threshold (F_6^{ED})
- plastic strain (γ_6^p) as a function of total applied strain (γ_6)

Cyclic shear stress–strain tests are used to obtain the nonlinear damaging behavior $\sigma_6(\gamma_6)$, as shown in Figure 9.2. The loading modulus is measured within a range of strain specified by the standard. The unloading modulus is measured over the entire unloading portion of the data.

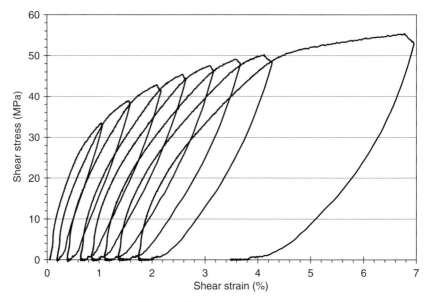

Fig. 9.2 Shear stress–strain behavior of unidirectional, neat specimen (no self-healing system). Loss of stiffness and accumulation of plastic strain are evident [3].

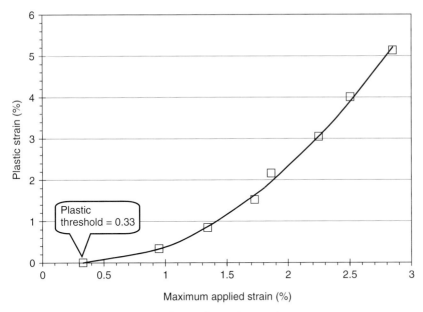

Fig. 9.3 Plastic strain versus applied strain for unidirectional, neat specimen (no self-healing system). Threshold plastic strain (i.e. yield strain) is evident [3].

Unrecoverable (plastic) strain can be observed upon unloading, but only after a threshold value of stress (i.e. the yield strength) or strain (i.e. the yield strain) is reached during loading (Figure 9.3).

Even though plastic strain is accumulated, initially the unloading modulus remains unchanged and equal to the loading modulus. For the unloading modulus to change, that is, to decrease below the value of the initial loading modulus, damage must appear. Note that the word "unloading" is added for emphasis and because the reduction in modulus is first detected during unloading of the specimen. But, of course, the modulus reduction is permanent. A reduction in the unloading modulus with respect to the initial loading modulus can be observed only after a threshold value of stress (i.e. the damage threshold stress) or strain (i.e. the damage threshold strain) is reached during loading (Figure 9.4).

The threshold stress F_{6EP} for appearance of unrecoverable (plastic) strain (i.e. the yield strength) and the threshold stress F_{6ED} for appearance of irreversible damage are read from the loading portion of the $\sigma_6(\gamma_6)$ curve (Figure 9.2) with the aid of Figures 9.3 and 9.4 (see also [5, chapter 9]). The unrecoverable (plastic) strain γ_6^p as a function of the applied total strain γ_6 is read for each cycle after full unloading and reported in Figure 9.3. The slope of the unloading curves provides the damaged elastic modulus G_{12}^d as a function of total applied strain γ_6 (Figure 9.4).

Existence of damage and a damage threshold are demonstrated by the fact that measured unloading modulus is less than the loading modulus after the damage threshold has been reached (Figure 9.4). No loss of stiffness occurs when the

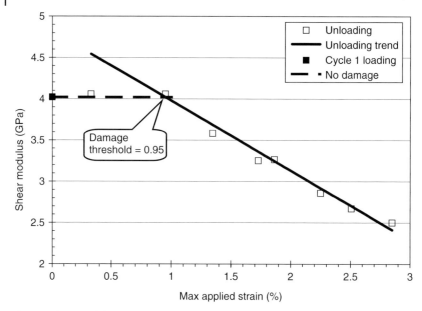

Fig. 9.4 Shear modulus versus applied strain of unidirectional, neat specimen (no self-healing system). Threshold damage strain is evident [3].

applied strain is less than the threshold. After the threshold is reached, the loss of modulus is proportional to the applied strain. Since careful visual inspection after each loading cycle does not reveal appearance of any macrocrack, the loss of modulus is attributed and modeled as distributed damage.

Also noticeable in Figure 9.3 is the accumulation of unrecoverable (plastic) strain. Although the physical, microstructural, and morphological mechanisms leading to plasticity in polymers are different than those leading to plasticity in metals, from a phenomenological and modeling point of view, unrecoverable deformations can be modeled with plasticity theory as long as the plastic strains are not associated to a reduction in the unloading modulus. The reduction in unloading modulus, which occurs independent of the plastic strain, can be accounted for by continuum damage mechanics. Each of these two phenomena has different thresholds for initiation and evolve with different rates. They are, however, coupled by the redistribution of stress that both phenomena induce. In the model, this is taken into account by formulating the plasticity model in terms of effective stress computed by the damage model [4].

Shear tests reveal marked nonlinearity (Figure 9.2) reaching almost total loss of tangent stiffness prior to failure, which occurs at large values of shear strain. Unloading secant stiffness reveals marked loss of stiffness due to damage, which worsens during cyclic reloading (Figure 9.4). Also, unloading reveals significant plastic strains accumulating during cyclic reloading (Figure 9.3).

The standard test method ASTM-D-3039 is used to determine E_1, E_2, v_{12}, F_{1t}, F_{2t}. The standard test method SACMA-SRM-1R-94 is used to determine F_{1c}, F_{2c}. The standard test method ASTM-D-5379 is used to determine G_{12}, F_6. The configuration of ASTM-D-5379 is used to determine G_{12} as a function of damage, healing, and number of cycles. Also γ_6^p, F_6^{EP}, F_6^{ED} are found using the same test configuration. The identification procedure linking the damage and plasticity parameters to the measured material properties is described in Ref. [5, Chapter 9].

9.5
Healing Identification

For healing modeling, all that is required is experimental determination of the healing efficiency as a function of damage. Under shear loading G_{12}, the amount of damage d_1 in the fiber direction is negligible when compared to the amount of damage d_2 transverse to the fibers. Therefore, the change in the (unloading) shear modulus due to both damage and healing is given by

$$G_{12}^d = \overline{G}_{12}(1 - d_2 + h_2) \tag{9.22}$$

and the healing efficiency can be calculated as

$$\eta_2 = \frac{G_{12}^h - G_{12}^d}{\overline{G}_{12} - G_{12}^d} \tag{9.23}$$

Taking into account that the induced damage d_2 is a function of the applied strain, it is possible to represent the efficiency as a function of damage with a polynomial as follows:

$$\eta_2 = 1 + a\, d_2 + b\, d_2^2 \tag{9.24}$$

as shown in Figure 9.5.

Twenty-two unidirectional samples containing self-healing system were loaded in shear with one and one-half cycles consisting of loading, unloading, followed by 48 h of healing time, and reloading [3]. Each specimen was loaded to a unique value of maximum applied shear strain in the range 0.5–4.0% with roughly equal number of specimens loaded up to 0.5, 1.0,..., 4.0% at intervals of 0.5%. A yield strain threshold value of 0.43% was found for the specimens with self-healing system [3].

Recovery was measured for each level of strain and from it the healing is calculated using Equation 9.23. The tests used to characterize efficiency consist of two loading cycles separated by healing. For the particular situation of just one reloading after healing, such as in these tests, efficiency is a function of strain as well as damage. However, for the more general case of multiple loading cycles, the amount of damage is a more appropriate independent variable to define the efficiency function.

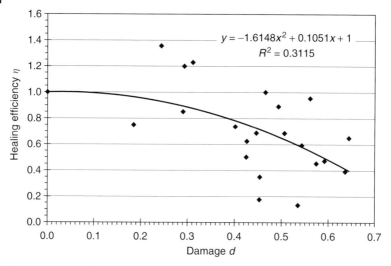

Fig. 9.5 Healing efficiency versus transverse damage.

Damage is a state variable; that is, damage describes univocally the state of accumulated damage regardless of the path followed to reach such damage. The total strain applied during multiple loading cycles is not a state variable because it can be achieved by various combinations of strain applied on each cycle. Each different combination would yield, in general, a different amount of damage. Therefore, even though applied strain is measured in the experiments, the amount of induced damage, not strain, is used to define the efficiency function. Shear tests are performed

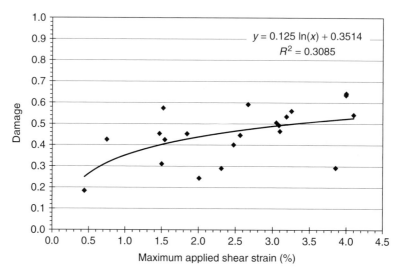

Fig. 9.6 Owing to hardening, increasingly large amounts of strain must be applied in order to produce more damage.

at strain levels of 0.25...4%. The healing efficiency from all specimens at each strain level (or damage level) are used to fit Equation 9.24, as shown in Figure 9.6.

9.6
Damage and Healing Hardening

Damage is represented by a state variable that accounts for the history of damage along the material principal directions (1, fiber-; 2, transverse-; 3, thickness direction). Once a certain level of damage is present, it takes higher stress (or strain) to produce additional damage (Figure 9.6). In this case, it is said that the material hardens. Damage hardening is represented by a hardening function, which is a function of the hardening variable δ. The effect of the hardening function is to enlarge the damage surface (Equation 9.12) that limits the stress space where damage does not occur. Damage hardening is represented in Figure 9.6 by a logarithmic function

$$d_2 = a' \ln(\gamma) + b' \tag{9.25}$$

as shown in Figure 9.5. It can be seen that additional strain must be applied in order to increase the amount of damage, thus hardening takes place.

Healing has two effects. First, it reverses some or all of the damage so that the stiffness of the material is recovered. At the same time, it resets the hardening threshold to a lower value, so that new damage can occur upon reloading at lower stress than would otherwise be necessary to cause additional damage on an unhealed material. This is merely a computational description of experimental observations. Such behavior can be interpreted as follows. The healed material can be damaged by reopening of the healed cracks, by creating new cracks, or by a combination thereof. In any case, the hardening function must be reset upon healing to be able to represent correctly the observed behavior.

In order to update hardening due to healing, first it is necessary to calculate how much of the damage can be healed. Since the self-healing system can only heal matrix damage, the damage that can be healed in the fiber direction is zero. The total damage that can be healed is then calculated as the sum of the damage in the two directions that can be healed

$$d^h = d_2 + d_3 \tag{9.26}$$

Next, the ratio of damage that can be healed in each direction to the total damage are calculated as

$$d_2^h = \frac{d_2}{d^h}; \quad d_3^h = \frac{d_3}{d^h} \tag{9.27}$$

By taking the healing efficiency into account, the amount of hardening recovered from healing in each direction can be calculated as

$$\mu_2 = \eta_2 \delta; \quad \mu_3 = \eta_3 \delta \tag{9.28}$$

Then, the overall hardening recovered from healing is calculated as

$$\mu = \mu_2 d_2^h + \mu_3 d_3^h \tag{9.29}$$

The amount of hardening recovered μ due to healing depends on the amount of healing that occurs in each direction. If the damage and healing phenomena are dominant in one direction, then that direction's healing efficiency will control the amount of recovery of hardening. Finally, the damage hardening parameter is updated as

$$\delta = \delta - \mu \tag{9.30}$$

9.7
Verification

The damage-healing model was identified using experimental data from a [0/90] symmetric as explained in Sections 9.4–9.6. Material properties are shown in Table 9.1.

ANSYS is compiled with a user subroutine implementing the damage-healing model. Finite element analysis is then used to represent the behavior of the sample materials. Model prediction and experimental data from a [0/90] specimen not used in the material characterization study and a quasi-isotropic $[0/90/45/-45]_S$ laminate are presented here for verification.

Shear tests of a single $[0/90]_S$ specimen was preformed. Quasi-static tests of the specimen loaded to 2.25% strain, unloaded, healed, and loaded again are shown in Figure 9.7. It can be seen that the computational model tracks the damaging Stress–strain behavior very well. Furthermore, the healing efficiency for this particular specimen was calculated using Equation 9.23 and used in the model. Comparison between experimental data and model prediction for the second (healed) loading of the specimen is shown also in Figure 9.7, where it can be seen

Table 9.1 Material properties of unidirectional composite.

Material property	Without self-healing	Standard deviation	With self-healing	Standard deviation
E_1 (MPa)	34 784	2185.89	30 571	4185
E_2 (MPa)	13 469	587.32	8699	829
ν_{12}	0.255	0.032	0.251	0.035
G_{12} (MPa)	3043	439.74	2547	207
F_{1t} (MPa)	592.3	29.32	397	66
F_{1c} (MPa)	459.1	43.66	232	59
F_{2t} (MPa)	68.86	9.17	45	10
F_{2c} (MPa)	109.5	9.25	109	9
F_6 (MPa)	49.87	3.39	38	2

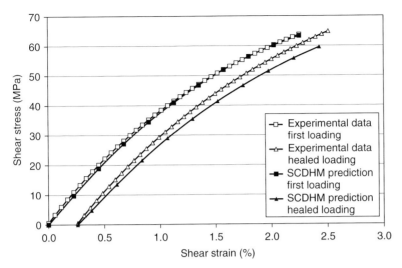

Fig. 9.7 Comparison between predicted response and experimental data for the first loading (damaging) and second loading (after healing) of a [0/90]$_S$ laminate.

that the computational model tracks reasonably well the damaging stress–strain behavior after healing.

Shear tests of a quasi-isotropic $[0/90/45/ - 45]_S$ laminate were preformed. Damage-healing tests of three specimens loaded to 1.5% strain and four specimens loaded to 2.25% strain were conducted. The loading shear stress–strain data of each specimen is then fitted with the following equation:

$$\sigma_6 = a'' + b'' \exp(-k'' \gamma_6) \tag{9.31}$$

The parameters a'', b'', k'' of all the specimens loaded to the same strain level (say 2.25%) are then averaged. Comparison of model predictions with the first (damaging) loading up to 2.25% strain is shown in Figure 9.8. The damage model predicts the actual damaging behavior very well. This is notable because the model parameters were adjusted with an entirely different set of samples, which shows significant variability (see Table 9.1 and [3, Table 2]). Comparison of model predictions with the second loading (after healing) of the same set of four samples is shown in Figure 9.8. Again, the accuracy of the model is remarkable.

In summary, microcapsules were fabricated in the same manner outlined in the literature. Grubbs' first generation ruthenium catalyst was encapsulated in the same manner outlined in the literature. Fiber-reinforced laminates were fabricated with the self-healing system dispersed within the laminae. Tests were conducted to quantify the damage, plasticity, and healing parameters in the self-healing computational model. Additional tests on samples not used in the parameter identification were performed in order to verify the predictive capabilities of the

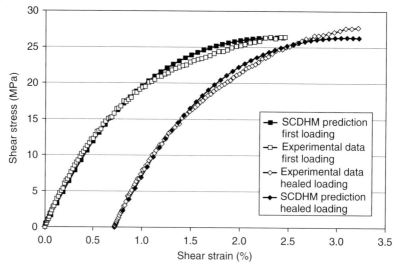

Fig. 9.8 Comparison between predicted response and experimental data for the first loading (damaging) and second loading (after healing) of a $[0/90/45/ - 45]_S$ laminate.

proposed model. It is observed that the proposed computational model tracks well the loss of stiffness due to damage, damage hardening, healing recovery, healing hardening, and damaging stress–strain behavior after healing.

References

1 Barbero, E.J. (**1999**) *Introduction to Composite Materials Design.* Taylor & Francis, Philadelphia, PA.

2 White, S., Sottos, N., Geubelle, P., Moore, J., Kessler, M., Sriram, S., Brown, E. and Viswanathan, S. (**2001**) "Autonomic healing of polymer composites". *Nature*, **409** 794–97. (Erratum *Nature*, **415** (6873), 817).

3 Barbero, E.J. and Ford, K.J. (**2007**) "Characterization of self-healing fiber-reinforced polymer-matrix composite with distributed damage". *Journal of Advanced Materials (SAMPE)*, **39** (4), 20–27

4 Barbero, E.J., Greco, F. and Lonetti, P. (**2005**) "Continuum damage-healing mechanics with application to self-healing composites". *International Journal of Damage Mechanics*, **14** (1), 51–81.

5 Barbero, E.J. (**2007**) *Finite Element Analysis of Composite Materials*, Taylor & Francis, Boca Raton, FL.

6 White, S., Sottos, N., Guebelle, P., Moore, J., Kessler, M., Sriram, S., Brown, E. and Viswanathan, S. (**2001**) "Autonomic healing of polymer composites". *Nature*, **409** (6822), 794–97.

7 Brown, E., Sottos, N. and White, S. (**2002**) "Fracture testing of a self-healing polymer composite". *Experimental Mechanics*, **42** (4), 372–79.

8 Kessler, M. and White, S. (**2001**) "Self-activated healing of delamination damage in woven composites". *Composites Part A Applied Science and Manufacturing*, **32** (5), 683–99.

9 Miao, S., Wang, M.L. and Schreyer, H.L. (**1995**) "Constitutive models for healing of materials with application to compaction of crushed

rock salt". *Journal of Engineering Mechanics*, **121** (10), 1122–29.

10 Jacobsen, S., Marchand, J. and Boisvert, L. (**1996**) "Effect of cracking and healing on chloride transport in opc concrete". *Cement and Concrete Research*, **26** (6), 869–81.

11 Jacobsen, S. and Sellevold, E.J. (**1996**) "Self healing of high strength concrete after deterioration by freeze/thaw". *Cement and Concrete Research*, **26** (1), 55–62.

12 Nakao, W., Mori, S., Nakamura, J., Takahashi, K., Ando, K. and Yokouchi, M. (**2006**) "Self-crack-healing behavior of mullite/sic particle/sic whisker multi-composites and potential use for ceramic springs". *Journal of the American Ceramic Society*, **89** (4), 1352–57.

13 Ando, K., Furusawa, K., Takahashi, K. and Sato, S. (**2005**) "Crack-healing ability of structural ceramics and a new methodology to guarantee the structural integrity using the ability and proof-test". *Journal of the European Ceramic Society*, **25** (5), 549–58.

14 Adam, J. (**2000**) A mathematical model of wound healing in bone, in *Proceedings of the International Conference on Mathematics and Engineering Techniques in Medicine and Biological Sciences*. METMBS'00, Vol. **1**, CSREA Press University of Georgia, Las Vegas, NV, pp. 97–103.

15 Adam, J. (**1999**) "A simplified model of wound healing (with particular reference to the critical size defect". *Mathematical and Computer Modelling*, **30** (5-6), 23–32.

16 Simpson, A., Gardner, T.N., Evans, M. and Kenwright, J. (**2000**) "Stiffness, strength and healing assessment in different bone fractures - a simple mathematical model". *Injury*, **31**, 777–81.

17 Lubliner, J. (**1990**) *Plasticity Theory*, Macmillan, Collier Macmillan, New York.

18 Schwartz, M.M. (**1997**) *Composite Materials: Properties, Non-Destructive Testing, and Repair*. Prentice Hall PTR, Upper Saddle River, NJ, Vol. **1**.

19 Wool, R. and O'Connor, K. (**1981**) "A theory of crack healing in polymers". *Journal of Applied Physics*, **52** (10), 5953–63.

Index

Self-healing Materials: Fundamentals, Design Strategies, and Applications. Edited by Swapan Kumar Ghosh
Copyright © 2009 WILEY-VCH Verlag GmbH & Co. KGaA, Weinheim
ISBN: 978-3-527-31829-2